MOTHS
OF THE WORLD

MOTHS OF THE WORLD

A NATURAL HISTORY

David L. Wagner

PRINCETON UNIVERSITY PRESS
PRINCETON AND OXFORD

Published in 2025 by Princeton University Press
41 William Street, Princeton, New Jersey 08540
99 Banbury Road, Oxford OX2 6JX
press.princeton.edu

Copyright © 2025 by Quarto Publishing plc

All rights reserved. No part of this publication may be reproduced or transmitted in any form, or by any means, electronic or mechanical, including photocopying, recording or by any information storage-and-retrieval system, without written permission from the copyright holder. Requests for permission to reproduce material from this work should be sent to permissions@press.princeton.edu

Library of Congress Control Number: 2024930903
ISBN: 978-0-691-24828-8
Ebook ISBN: 978-0-691-24829-5
British Library Cataloging-in-Publication Data is available

This book was conceived, designed, and produced by
The Bright Press, an imprint of the Quarto Group
1 Triptych Place, 2nd Floor,
London SE1 9SH, United Kingdom
www.Quarto.com

Publisher **James Evans**
Editorial Directors **Isheeta Mustafi, Anna Southgate**
Managing Editor **Jacqui Sayers**
Art Director and Cover Design **James Lawrence**
Project Manager **David Price-Goodfellow**
Design **Ginny Zeal**
Illustrations **Robert Brandt** (diagrams) **and Les Hunt** (maps)
Picture Research **Julia Ruxton**

Cover Photos: Front cover: (left column, top to bottom) Piotr Naskrecki / Piotr Naskrecki / Shutterstock: Filip Fuxa / David L. Wagner / Flickr: Andy Reago and Chrissy McClarren / Shutterstock: Mikhail Melnikov / Piotr Naskrecki; (center column, top to bottom) iNaturalist: Lucy Keith-Diagne / iNaturalist: Christoph Moning / Shutterstock: Simia Attentive / (left) Flickr: Bernard Dupont / (right) Piotr Naskrecki; (right column, top to bottom) Shutterstock: Danut Vieru / Shutterstock: Melinda Fawver / Piotr Naskrecki / Flickr: Janet Graham / iNaturalist: Kahio T Mazon / Shutterstock: Protasov AN / Shutterstock: JPS; Back cover: Flickr: Alexey Yakovlev; Spine: Shutterstock: SIMON SHIM.

Printed and bound in Malaysia
10 9 8 7 6 5 4 3 2 1

Dedication
This book is dedicated to my parents, George and Kaye, and my family, Sylvia, Virginia, and Ryan; the mentors who educated and inspired: Alexander Barrett Klots, F. Brent Reeves, Jerry A. Powell, Fredrick W. Stehr, and Howard Ensign Evans; and Richard P. Garmany, a champion of nature, who has supported my work for two decades.

6 Introduction

CONTENTS

82 Non-ditrysian Microlepidoptera

102 Ditrysian Microlepidoptera

164 Macrolepidoptera

231 Epilogue: The Sixth Extinction, Nature, and Moths
232 Glossary
234 Additional Reading and Important Resources
235 Index
240 Acknowledgments and Picture Credits

INTRODUCTION

Whether measured by abundance, biomass, or numbers of species, moths are one of the most successful branches on the tree of life. Moths are the fabric of many forests, woodlands, and shrublands ecosystems. By consuming plants and later being consumed themselves—by insectivores such as frogs, lizards, snakes, birds, rodents, skunks, foxes, bats, monkeys, and even bears—they transfer more nutrients and energy from plants to animals than any other group of insects. Spring would be silent indeed without moths.

While the lion's share of moths lack flamboyant coloration—color is less important in nocturnal animals—their ranks include some of the world's most beautiful invertebrates. Few creatures can rival the radiance of a Madagascan Sunset Moth, or the subtle beauty of a freshly emerged Luna Moth. The caterpillars of butterflies and moths are extraordinary animals of great ecological importance and seemingly infinite in their different forms and strategies for avoiding predation.

Given the many threats of the Anthropocene to nature, there is a great and immediate need to learn more about these denizens of the night, to document their value, and to accelerate education and conservation efforts that will ensure their

survival. Knowledge of their early stages and life histories is essential to any conservation plan seeking to protect an imperiled insect species. For students of nature, moths offer boundless opportunities for study and discovery. In part, this is why moths and their caterpillars are enjoying rapidly increasing interest among photographers, community scientists, ecologists, and land managers. This book will introduce you to their rich diversity and variety, their ecological importance, key anatomical attributes, and biology. It will also convey much about what is unknown and remains to be discovered.

ABOVE LEFT | *Parasa indetermina* is wonderfully varied in color, with most rendered in yellows, oranges, or reds that warn against touching, as the caterpillar is armed with batteries of stinging spines.

ABOVE | Mating Luna Moths (*Actias luna*), one of North America's most beloved invertebrates. Note the difference in the development of the antennae across the two sexes.

DIVERSITY, CLASSIFICATION, AND EVOLUTIONARY HISTORY

Moths and butterflies comprise the order Lepidoptera, which in Greek means "scaled wings." More than 160,000 species have been described, with likely two to three times that number yet to be named, mostly from the tropics. While Lepidoptera are currently the second-largest order of animals, after beetles (Coleoptera), recent molecular studies suggest both orders will be greatly exceeded by gnats, midges, and flies (Diptera) and the order for ants, wasps, bees, and kin (Hymenoptera). In the northern hemisphere, where moth faunas are better known, butterflies comprise about 6 percent of the lepidopteran fauna. Stated differently, for every butterfly species, there are at least 15 times as many species of moths, with most of these becoming active only after we have gone into our homes for the night.

DIVERSITY

Moth diversity peaks on the east side of the Andes and across the adjacent mountain ranges of

BELOW | Infinitely varied in color and form, moth scales serve important roles in defense, communication, thermoregulation, and still other functions.

western South America. While the area around Tiputini, Ecuador, may have the single greatest species richness per hectare, Colombia may prove to have the single highest total, given its large area and the fact that its border includes three faunally distinct mountain ranges. More than 2,500 butterfly species have been documented along a single road that drops from the Andean Highlands of Peru down into the Amazon basin. If the ratio of butterflies to moths in the northern hemisphere holds for Peru, we can expect this same road to yield some 39,000 species of moths. That is more than twice the number known to occur in North America north of Mexico, and a quarter of all presently described species.

While exceptions are many, species diversity decreases in higher latitudes as abiotic conditions, especially winter temperatures, get increasingly limiting. On many mountains, diversity peaks in the foothills and low elevations. Diversity drops off in deserts and other arid habitats. As might be expected, moth diversity is closely linked to plant diversity, and especially that of perennial angiosperms (flowering plants)—trees host more species than shrubs, which in turn host more species than herbaceous perennials, which harbor more species than annual plants. Plant taxa that are diverse in kind and ecologically dominant—that is, those that represent large "biological islands"—such as oaks in the north temperate zone and acacias and figs in the tropics—are food for far more species of moth caterpillars than small genera and plants that are taxonomically isolated.

BELOW | A good moth night at a sheet sited between two mercury vapor lights in Sonora, Mexico. Seasonally dry forests—away from anthropogenic stressors—can explode with moths following the arrival of monsoonal rains.

PHYLOGENY AND GEOLOGICAL HISTORY

The oldest confirmed moth fossil dates back to about 212 million years ago (mya), the Late Triassic, when dinosaurs reigned over land, sea, and air; DNA data suggest the order had a much earlier origin, extending back into the Permian. The oldest surviving family of the jawed moths (Micropterigidae) is represented by more than 260 species with many of these from the southern hemisphere, especially New Caledonia, which is widely recognized for its relict flora and fauna.

The evolutionary tree (opposite) for the order emphasizes the lineages that survive today and is largely blind to the many lineages that went extinct, as we have relatively few moth fossils that can tell us about the order's diversity over much of its history. It is thought that moths were in existence for 80–120 million years after their origin, budding off new families of modest ecological importance until about 125 mya, in the Cretaceous, when the explosive radiation of angiosperms provided a multitude of new niches for moths and their caterpillars.

The 24 surviving relict families are informally referred to in this guide as the archaic taxa. Many of the modern families of Lepidoptera appear to have had their origins in the period between 120–100 mya, when flowering plants were radiating. It is likely that moths played important roles in the rapid diversification of angiosperms, both as pollinators (the adults) but also as plant enemies (the larvae). The informal category of microlepidoptera refers to the moth lineages up through pyraloids and mimallonoids; the Lasiocampoidea and upward include the moth taxa commonly referred to as macrolepidoptera. Evolutionarily speaking, butterflies are moths! They represent no more than a couple of origins of brightly colored, diurnal (day-flying) moths, in the same way that birds are just a specialized lineage of dinosaurs. For this reason, butterflies are mentioned at many points in this volume and are covered in full in another book in this series.

LEFT | *Sabatinca kristenseni*, a particularly handsome micropterigid, only 0.3 in (7 mm) or so in length, is thought to be a jumping spider mimic, which serves to protect it from similarly sized and smaller spiders.

RIGHT | Phylogenetic tree of major moth lineages adapted from Kawahara *et al.* (2019). (Mimallonidae and other taxa are treated in the phyologenies added to the taxon profile introductions.) Tineoidea appears twice as they are not monophyletic, that is they fall into at least two separate groupings. This tree is provisional; for example, other data support Sesioidea being nested within Cossoidea.

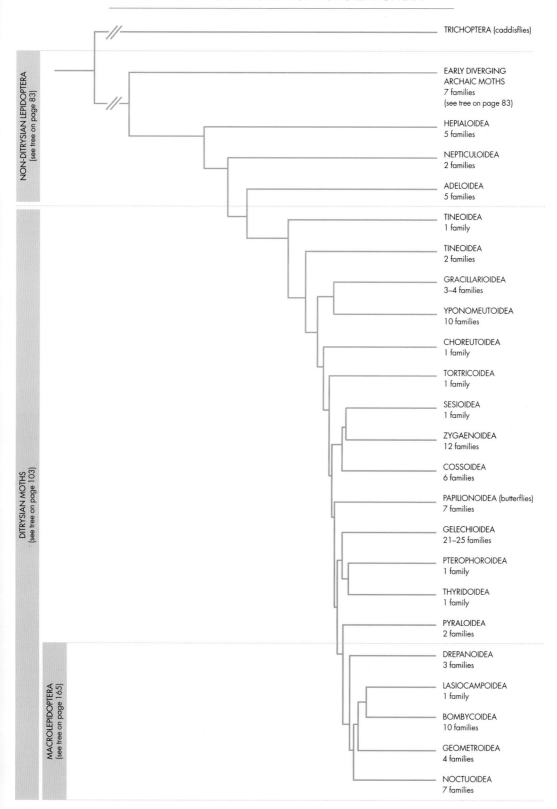

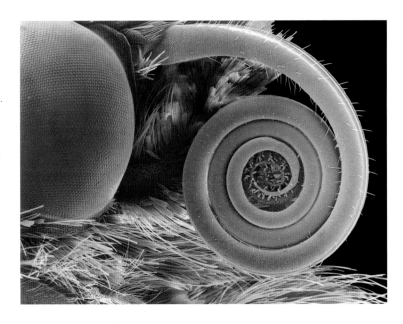

RIGHT | Colorized scanning electron micrograph (SEM) of the extensible tongue of a lepidopteran. This haustellum is heralded as a key adaptation of the order, allowing moths to access water and food resources unavailable to other insects.

BELOW | While most micropterigids will be encountered near the ferns and liverworts on which they feed, many of the more than 75 species of *Micropterix* often visit flowers. This is *Micropterix aureatella*.

The order Lepidoptera includes nearly 140 extant and four extinct families. Of the extant families, seven are butterflies. They are classified into four suborders: Zeugloptera, Aglossata, Heterobathmiina, and Glossata. The first three of these collectively include about 300 species. The Glossata are moths that have a coilable haustellum or tongue (see page 17), which allowed early moths to suck up nutrients and water. These, in turn, are parsed across three infraorders: Dacnonypha, Exoporia, and Heteroneura. The last of these, with wing venation that differs markedly between the fore- and hindwings, includes 99 percent of all described lepidopteran species. The collective set of "early diverging" taxa that originated before Heteroneura includes 14 superfamilies, 8 of which are profiled in this book. With the exception of the Hepialidae (ghost moths), these relict lineages tend to be small insects; most are uncommon to extremely rare, with some species known only from a single or very few individuals. Adults of several of these archaic lineages have metallic gold and purple forewings or are otherwise quite handsomely patterned. The Hepialidae are medium to large moths; most are drably rendered in earth tones, but a few are exquisite insects, and include some of nature's most beautiful creatures. All archaic moths make good quarry for enthusiasts and represent an opportunity to see what moths might have looked like and how they behaved in the Mesozoic when dinosaurs were at the peak of their reign.

Heteroneura are divided into seven clades parsed across 35 superfamilies and 125 families (including 7 butterfly families). We introduce all but three of the 42 moth superfamilies; 72 families and 29 subfamilies are accorded a separate taxon profile. Much moth diversity, close to 98 percent of all described species, is housed by a single clade, the Ditrysia. Ditrysian females have separate reproductive openings for copulation and egg-laying; the two are connected by an internal duct.

Ditrysia includes 29 superfamilies and 110 families—and accounts for nearly all the moths the average person will take note of over the course of their lifetime. Just five of these families account for half of all described moths: in decreasing order of species (using the last global tally published in 2011): Erebidae (24,569), Geometridae (23,002 species), Noctuidae (11,772), Tortricidae (10,387), and Crambidae (9,655). The three largest families are macrolepidopterans with primarily externally feeding larvae—that is, their caterpillars tend to feed exposed on leaves and other substrates. Four of these families, but not the Tortricidae, have a well-developed ear tuned to detect the ultrasonic hunting frequencies of feeding bats. Moreover, three separate evolutionary origins of an ear are represented: one in Noctuoidea (Erebidae and Noctuidae), one in Geometroidea, and another in Pyraloidea—collectively underscoring the quintessential importance of having a bat

BELOW | Externally feeding caterpillars employ a battery of strategies to foil their natural enemies. This notodontid (*Ianassa lignicolor*) mimics the abandoned, necrotic leaf folds formed by gall midges in the spring.

detector to the diversification and ecological successes of the order.

The phylogeny shown on page 11 is greatly simplified, showing only some of the major higher categories of moths. A phylogeny for all 134 families would show a dramatic evolutionary explosion of families when angiosperms (flowering plants) became the dominant life forms across Earth. So swift was the concomitant radiation of moths that there remains uncertainty about the origins and phylogenetic relationships of several moth families that evolved during this time of rapid diversification.

TAXONOMIC NAMES

Four taxonomic ranks anchor the text in this work: order, family, genus, and species. The first two of these are "higher categories" to which prefixes can be added to indicate subordination or rank. Commonly you will see super-, sub-, infrasub- used to indicate ranking about a given higher category. A special term adopted in the work is clade to gather moths that appear to share a common ancestor within the infraorder Heteroneura. Taxon (pl. taxa) is used to refer to any taxonomic entity on the tree of life, without having to specify its rank, and thus could be used for a species, group of species, on up to a group of families or orders.

Higher-category names above the family level do not have standardized endings indicating their rank. Latin suffixes for family-group names are standardized across all zoological classifications (see table). Common or vernacular names based on the Latin higher-category may be made by adding English suffixes. The first letter of higher-category Latin names must be capitalized (Acronictinae); capitalization is optional for the common name based on this name (Acronictines or acronictines). Genus and species name are *italicized* by convention.

Category	Latin suffix (example)	Latinized suffix common name (example)	English common name
Superfamily	-oidea (Noctuoidea)	-oids (noctuoids)	
Family	-idae (Noctuidae)	-ids (noctuids)	owlets
Subfamily	-inae (Noctuinae)	-ines (noctuines)	e.g. cutworms, sallows
Tribe	-ini (Orthosini)	-ini (orthosines)	quakers and woodlings

While higher-category names above the superfamily level are unstandardized and never italicized, they still must be capitalized. They are vernacularized by placing an "n" as the suffix: for example, Lepidoptera and lepidopterans are synonymous, as are Heteroneura and heteroneurans.

WHY SO MANY MOTHS?

Lepidoptera comprise one of the most speciose and ecologically important groups of animals on our planet. In many regions of the world, moths are among the most conspicuous insects in yards: as caterpillars in gardens, adults visiting flowers, and the insect life spiraling about our porch lights. In tropical forests and some grasslands, moths and their caterpillars are thought to transfer more energy from plants to other organisms than any other plant-feeding animal group. Caterpillars account for much of the diet for the world's songbirds—that alone underscores that they are among the more ecologically important lineages linking temperate and tropical ecosystems. This section is about why this might be.

The answer may be no more complicated than that 99 percent of all moths are plant feeding. Moths flourished as angiosperm plants diversified, beginning in the early Cretaceous 125 mya. The 370,000 recognized species of vascularized plants make the lands of the world green and have provided endless possibilities for the diversification of moths. Indeed, moths and butterflies comprise the largest single radiation of plant feeders on the planet—with the fates of Lepidoptera and plants now inextricably intertwined.

COMPLETE METAMORPHOSIS AND WINGS

Two traits common to the planet's most megadiverse insect orders—the beetles

ABOVE | Caterpillars are a mainstay in the diets of many birds and a critical resource for raising nestlings.

RIGHT | Across many of the planet's forests and woodlands, moths and their caterpillars are essential elements of food webs.

(Coleoptera); flies (Diptera); bees, ants, and wasps (Hymenoptera); and moths and butterflies (Lepidoptera)— account for much of the success of these orders: their holometabolous development and wings. Holometabolous insects have a distinct larval stage that differs from the adult stage to such an extent that the two are different creatures with unique morphologies and ecologies. Matching one with the other takes careful investigation. This mastery of gene regulation allows the larva to be a feeding machine with a singular mission: to eat and not be eaten. The larval stage, shaped by millions of years of evolution, has specialized mouthparts, digestive tract adaptations, and body form that allows it to feed, grow rapidly, and exploit resources unavailable to the adult. Likewise, the adult is fitted with wholly different mouthparts suited to still other resources (for example, nectar and other liquid substrates), a reduced gut, wings, genitalia, and all the other anatomical paraphernalia needed to reproduce successfully.

The importance of wings in the success of insects is obvious and undebatable. Early evolving orders without wings account for less than 1 percent of insect diversity, and lineages that have secondarily lost their wings tend to be relatively small taxa. Even more telling, most of these bear the earmarks of evolutionary recency. Stated differently, extinction lies in wait for those that forego the ability to fly, as they are often less well equipped to deal with natural enemies and environmental change. Wings allow insects to disperse and migrate, locate widely scattered resources, evade predators, seek and court mates, and defend territories. Scale colors and patterns of the wings play additional roles in the success of moths; foremost of these is predator avoidance. An enormous variety of male androconia (scent scales), essential to precopulatory communication,

LEFT | Burnet (*Zygaena*) moth sipping nectar. Many burnets—protected by toxic proteins and cyanide-yielding metabolites (β-cyanoalanine)— advertise their unpalatability with bright red wing spots.

ABOVE | Montane and high-latitude moths, such as *Macaria carbonaria*, are often diurnal and set with dark scales that absorb sunlight to warm the body.

are found on moth wings. In diurnal moths, wing patterns can be important in species recognition and during courtship displays.

SCALES

The wing scales of Lepidoptera have been important to the diversification and ecological success of moths. They serve in courtship and communication, and thus in sexual selection and maintaining species identities, as well as in thermoregulation, predator avoidance, and defensive chemistry. Moth scales, and especially the thin, hairlike scales of the body, function to absorb the pulsed calls of hunting bats, and thus serve as a cloaking device to thwart bat attacks. Their deciduous nature sometimes allows adult moths to escape entrapment in spiderwebs or slip out of the mouths of some vertebrate predators. Contrary to those warning that removing wing scales irreparably damages a moth or butterfly's ability to fly, know that the scales are not essential, but do contribute to an adult's flight-climbing efficiency. Indeed, many moths have much of the wing membrane free of scales. Scales that are laden with defensive chemicals provide a ready means by which a moth can warn a would-be enemy that it is chemically protected and best left alone. Montane and polar species often have dark scales over the thorax and along the wing veins; the latter trap heat, which is transferred into the thorax by the hemolymph.

COILABLE TONGUE

The most ancient moths had grinding mandibles that were soon lost by later evolving and more ecologically successful lineages that instead evolved haustella from modifications of the maxilla. Initially, these were short and likely functioned by wicking so that the adult could take up water or other pooled fluids. Over time, more elongated and biomechanically efficient haustella developed—tongues that could be coiled when not in use. Moths with a coilable haustellum account for more than 85 percent of extant

LEFT | Like hummingbirds, hawk or sphingid moths can fly forward and backward and hover while nectaring. The fully exerted tongue in a few species may extend more than 11 in (28 cm) in length.

BELOW LEFT | Male sunset moths "puddling." This behavior, common to many butterflies and moths, is typically driven by a desire to gather sodium, much of which is transferred to the female during copulation.

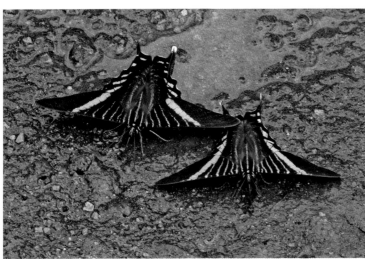

species. The haustellum allows the adults to feed at many fluids unavailable to many other insects—nectar and water, but also at the droppings of larger animals, fermenting fruits, decaying plant and animal matter, mud and moist soil, plant wounds, sweat, tears, and even blood. The first of these abilities, allowing for access to nectar, is the most important, and has made moths the preferred pollinators for many plants (see Pollination, page 46). The record holders here are the sphinx and hawk moths (Sphingidae).

The tongues of Wallace's Sphinx (*Xanthopan praedicta*) and the Giant Sphinx (*Cocytius antaeus*) sometimes exceed lengths of 11 in (28 cm).

NOCTURNALITY AND EARS

The archaic moths include both diurnal and nocturnal lineages; Agathiphagidae, Hepialidae, Nepticulidae, and others are believed to be principally night fliers. Looking across the tree of life, the vast majority of moth families, genera, and species are nocturnal. Until the time that bats became ecologically dominant some 50 mya in the Eocene, moths surely owned the night—a behavioral trait that has served them well and accounts for much of their eco-evolutionary success. Ears, tuned to the ultrasonic frequencies of feeding bats, have been quintessentially important to the ecological and evolutionary successes of moths. Structures sensitive to ultrasound have evolved independently in no fewer than six nocturnal moth lineages. With the exception of giant silk moths and a few others, nearly every diverse moth lineage with wingspans commonly exceeding ¾ in (2 cm) has evolved a bat detector. As a further testament: many of the moth lineages that reverted back to diurnality—butterflies and many geometrids—subsequently lost their ability to detect ultrasound.

LARVAL CRYPSIS

Caterpillars account for much aboveground insect biomass in terrestrial tropical, temperate, and boreal ecosystems. Stated differently, they are keystone organisms shaping life on land for insectivores: species of lizards and snakes; mammals from rodents to foxes, monkeys, and bears; most terrestrial birds; and an inestimable number of invertebrate predators. The vast majority of caterpillars are palatable—that is, to a large degree they lack chemical or physical defensive attributes necessary to discourage their predators. Caterpillars survive by avoiding detection; they are masters of camouflage, crypsis, and deceit. The ability of palatable species to background match, which can involve a battery of morphological and behavioral

BELOW | Crafted by millions of years of Darwinian natural selection, palatable caterpillars can be exceedingly cryptic in color, form, and habit.

TOP | Lichen-feeding noctuid caterpillar (*Bryolymnia viridata*) (positioned over the upper, central portion of the twig).

MIDDLE | Two twig-mimicking geometrid caterpillars (*Plagodis alcoolaria*).

BOTTOM | Seed-eating noctuid caterpillar (*Stiria intermixta*) that fills the cavity where it has fed with its sculptured body. All three of the animals on this page do most of their feeding at night—under cover of darkness.

ABOVE | Communal nest of fall webworm caterpillars excludes many would-be enemies: ants, spiders, wasps, and others.

solutions, represents a key adaptation that has allowed them to flourish, fully exposed on vegetation, to a greater extent than other insects.

SILK

The fibrous protein we call silk is one of the great evolutionary inventions of animal life. A testament to its importance is the number of independent origins: silk production has evolved in mussels, crustaceans, spider mites, sun or camel spiders, pseudoscorpions, and, of course, in spiders and repeatedly across at least a dozen insect orders and families. If measured either by species richness or ecological importance, spiders and caterpillars are the webmasters. Caterpillars use silk to make shelters and nests; these serve to exclude many natural enemies, prevent desiccation, or in the case of tent caterpillars serve both as a safe house and greenhouse that heats up well above ambient temperatures on cold spring days. More than a dozen moth families construct a portable silken case that conceals and protects the larva from natural enemies and desiccation. Cocoons are analogous constructions spun by prepupae to protect the pupal stages. Caterpillars belay themselves away from danger or seek a new feeding site on silk lines; early instars "balloon," dropping themselves on a short silk thread to disperse on air currents. Many geometrid caterpillars descend on a silk thread by night where they remain suspended until daybreak, to escape foliage-gleaning spiders, beetles, ants, and other enemies. Caterpillars spin molting pads, into which their crochets can be secured prior to a molt. Many caterpillars lay down silk trails that can be retraced: to shelter away from their host, to return to their host, or used by siblings to hold their cohort together.

EXTERNAL ANATOMY

Below we introduce externally visible structures of the adult and larval stages that are mentioned elsewhere in the book, especially those relevant to the recognition of higher-level taxa: superfamilies, families, and subfamilies.

ADULT

The 20 segments that make up a moth's body are divided into three parts: six fused segments that make up the head, three the thorax, and eleven the abdomen.

HEAD

The head bears four paired appendages, each corresponding to one of the ancestral segments: (from front to back) the antennae, mandibles, maxilla, and labium. The haustellum, or proboscis, is a derivative of the maxilla. The compound eyes and especially the optic lobes of the brain occupy much of the head volume. Many lepidopteran families also have ocelli, small light-processing organs between the compound eyes and above the antennae. The structures of the head most commonly used in moth identification include the structure of the antenna (especially in males), maxillary and labial palpi, and haustellum—both its length and whether or not it is scaled along its base.

Olfaction is a principal mode of communication in many animals that operate in darkness. Consequently, the antennae tend to be greatly elaborated in moths and often have species-specific features. In nearly all moths the male antennae are more branched, housing batteries of sensilla, specialized for detection of the female sex pheromone. Females typically have

EXTERNAL ANATOMY OF A MOTH

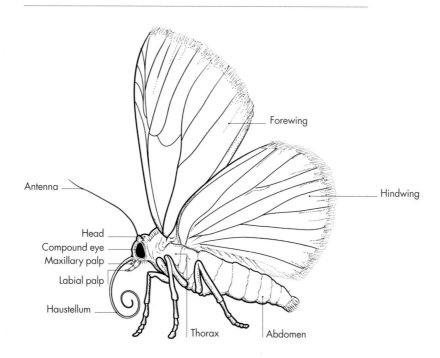

HEAD DETAIL

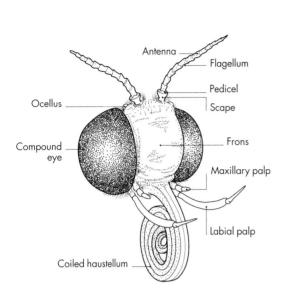

WING DETAILS AND VENATION

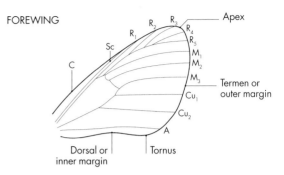

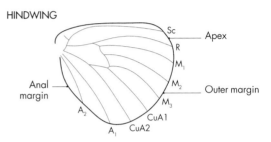

comparatively simple antennae used primarily to detect close-range male pheromones and volatile chemicals emanating from their host plants.

THORAX

The thorax is given to the business of locomotion. All three thoracic segments bear a set of legs, which differ in size, shape, and function. Each leg consists of four conspicuous parts: the coxa, femur, tibia, and subsegmented tarsus. The first thoracic segment, the prothorax, is a narrow collar-like segment. The two pairs of wings, forewings and hindwings, are borne from the last two highly muscularized segments that power

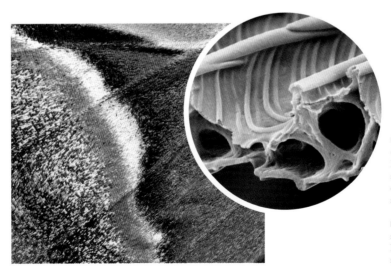

LEFT | Hindwing of Cecropia Moth (*Hyalophora cecropia*). The deciduous, hydrophobic nature of lepidopteran scales allows moths to sometimes escape from would-be predators, spiderwebs, and water films.

INSET | Cross-section of a moth scale. Ridges add strength, while the lacunae and hollow nature of the scales diminish each scale's mass.

ABOVE | The ear of noctuids sits in a thoracic cavity below the hindwing. The circular area in the center of the image is the tympanum. Note the somewhat whiter area in the central portion of the tympanum—this is the attachment site for the two master nerves that control the moth's evasive behaviors. See text below.

flight: the mesothorax and metathorax. Wing venation is used extensively by lepidopterists to characterize higher taxa and infer evolutionary relationships among taxa.

The foreleg is often modified in some way. In those moths that pupate underground, the foretibia often bears spurs, spines, or robust setae that are used for digging the moth's way to the soil surface. Across many families, there is a flap of cuticle, the epiphysis, on the inner side of the tibia, which serves as an antennal comb. Moths are able to groom each antenna by drawing it through the gap between the tibia and epiphysis. The hindlegs are the largest and most powerful of the three pairs, and often bear androconial scales on the femur or tibia that are displayed in precopulatory courtship displays.

Noctuoidea have an ear (cavity) below the hindwing with a thin membrane that vibrates when exposed to the high-frequency calls of hunting bats. There are two nerve cells connected to the thin tympanic membrane: one fires at the initial point that a bat is detected; the second neuron fires if the bat calls increase in frequency—a signal that the moth has been targeted by a hunting bat. This structurally simple ear, which allows noctuoid moths to detect and take evasive measures, is widely regarded to be a key adaptation that accounts for much of the singular success of the noctuoids, the most diverse superfamily of moths. The clade includes more than 44,000 described species—more than a quarter of all described species of moths.

The wing-coupling mechanism is also of taxonomic significance. In most extant lineages, there are one or more bristlelike spines (comprising the frenulum) that extend from the base of the hindwing and "lock" into a cuticular lobe or set of curved scales on the underside of the forewing. This keeps the wings locked into a single airfoil during flight. Across many families, males have a singular frenular bristle, while females have two or more slender frenular bristles. The most "primitive" or early diverging lineages of moths—for example, ghost moths—lack a frenulum and retinaculum: in these, the wings may move independently in flight as occurs in dragonflies and lacewings. In Bombycoidea, the base of the hindwing is greatly expanded and projects under the forewing, keeping the wings together when flying without a frenulum.

ABDOMEN

Ancestrally, insects had an 11-segmented abdomen; only nine are visible externally in adult moths, with the terminal segment assumed to include the remnants of segments 10 and 11. The pregenital abdomen, segments one to seven, is the epicenter of metabolism in insects, where much circulation, detoxification, digestion, excretion, fat storage, and respiration take place. Pyraloids and Geometridae have abdominal ears at the front of the abdomen, tucked in the cavity between the thorax and abdomen. Pregenital segments are commonly involved in pheromone production, and, in males, additionally house various scent brushes.

The principal genital segments are derived from segments eight and nine. In all but the most archaic lineages, the genitalia are internal, at least until the point of copulation. Structurally, the genitalia are morphologically complex in both sexes, but especially so in males. Not surprisingly, these and adjacent segments are often involved in pheromone production; male courtship brushes, made of scent-laden scales (androconia), are especially common on segment nine.

The male and female genitalia are of particular importance to moth taxonomy because they often have reliable species-level characters (see pages 50–52). In moth genera where the external features provide little clue as to identity— for example, in all white, all black, or otherwise uniformly colored taxa—countless cryptic moth species have been discovered upon dissection of the male or female genitalia. It is still an expectation that authors wishing to name a new species describe and illustrate the male and female genitalia.

LARVA

A caterpillar's body is organized in much the same fashion as the adult, with six fused head

BELOW | Most caterpillars have fleshy prolegs on abdominal (A) segments three to six (A3–A6) and a terminal pair on A10. (T = thorax.)

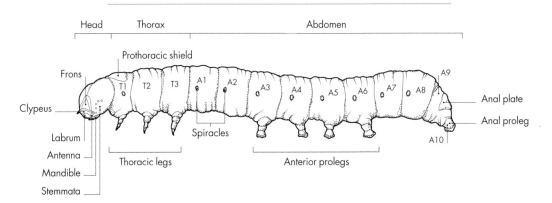

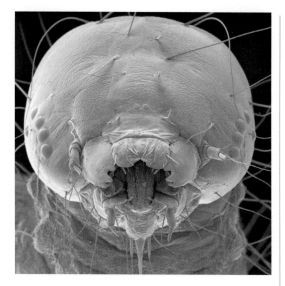

LEFT | Scanning electron micrograph showing the head of a metalmark moth. Above the long spinneret, the toothed mandibles are parted; lateral to the mandibles, note the small antennae and the series of stemmata.

segments, three thoracic segments, and ten distinguishable abdominal segments. The body is greatly simplified yet at the same time specialized for ingestion, digestion, and growth. Superficially a caterpillar appears to be little more than a hydrostatic tube with internal muscles, an enormous gut, and hypertrophied salivary (silk) glands.

HEAD

In contrast to those of an adult, larval antennae are simple structures, much reduced in size and function. The mandibles on the other hand are robust, hardened, and allow caterpillars to consume many tough substrates, including foliage, cones, fruits, and seeds and to bore into roots, stems, and wood. As in the adult, the maxillary and labial palpi, with important taste sensilla at their apices, play crucial roles in tasting (and rejecting) foods, and what gets brought to the oral cavity.

The spinneret, situated between the minute labial palpi, is little more than the external spigot of an enormous salivary-silk gland that stretches from the mouth back to the mid-abdomen—were the gland stretched linearly, its length would extend beyond the caterpillar's rump in some species. The gland is filled with semiliquid silk and a potpourri of salivary components that serve in digestion, immunity, and the inhibition of induced defenses of their host plants. Once secreted, the gland's contents solidify to yield the silk fibers that are used to fashion cocoons, nests, molting pads, belay lines, and more (see Silk, page 20).

TRUNK

The thoracic and abdominal segments are fused into a single trunk that houses an enormous gut for food storage and digestion. In addition to thoracic legs, caterpillars often bear fleshy prolegs on abdominal segments three to six and the last segment. At their apices, the prolegs bear a set of hard, recurved, cuticular hooks, the crochets. The arrangement of these—whether they form a circle, alternate in size, or occur in more than one row—is of great value in caterpillar identification.

Most caterpillars have eleven primary setae on each of the last two thoracic and first eight abdominal segments. Additions to this basic number are considered secondary setae. Both types of setae, their lengths, structure, and positions on the body wall differ among species. Caterpillars that feed externally on their food plant often can be identified by their coloration, in the same way that moths can be identified by their wing patterns. This is not the case for internally feeding species, which tend to be pale and undifferentiated. For these, the crochet arrangements and arrangement of bristles (chaetotaxy) provide the principal means by which entomologists identify caterpillars.

INTERNAL ANATOMY

Moths have organ systems that carry out the basic metabolic tasks demanded of all animals. Below, we emphasize those that differ fundamentally from those of vertebrates. Most obviously, insects have exoskeletons to which the muscles and connective tissues attach, which adds a significant constraint: their growth requires molting. Conversely, their outer "skeleton" has numerous advantages. Foremost among these, the integument provides protection from many predators, pathogens, and parasites. It also plays a fundamental role in preventing desiccation, and thus preadapted insects and their kin to be among the first creatures to crawl out of the oceans to invade terrestrial environments. For millions upon millions of years, insects and other terrestrial arthropods were able to rule land and sky before they were joined by amphibians and reptiles.

CIRCULATORY SYSTEM

Moths and other insects have a partially open circulatory system: the core body fluid, hemolymph, bathes all the living tissues of the body. An elongated dorsal heart and various auxiliary pumps found at the bases of the appendages—antennae, legs, wings—move the hemolymph through the body. Unlike vertebrate blood, hemolymph does not carry appreciable oxygen. Instead, respiration is achieved via the spiracles and internal tracheal system that delivers oxygen to every cell group of the body.

NERVOUS SYSTEM

The nervous systems of vertebrates and invertebrates, while similar and built from the same basic types of neurons, are organized quite differently. Vertebrates have a single dorsal nerve cord; that of insects is ventral and paired. More significantly, a unified brain is centralized in the head of vertebrates. The insect brain is tripartite and is followed by many semiautonomous ganglia in the region of the mouthparts and along the thoracic and abdominal segments. These ganglia, well rearward of the brain, can play important roles in sensory and metabolic functions, locomotion, mating, and hormonal signaling.

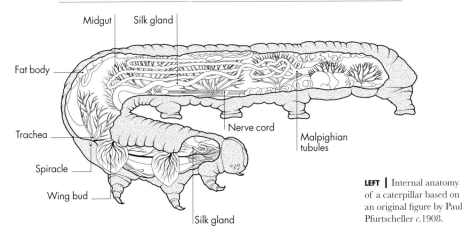

LEFT | Internal anatomy of a caterpillar based on an original figure by Paul Pfurtscheller *c.*1908.

LIFE CYCLE AND METAMORPHOSIS

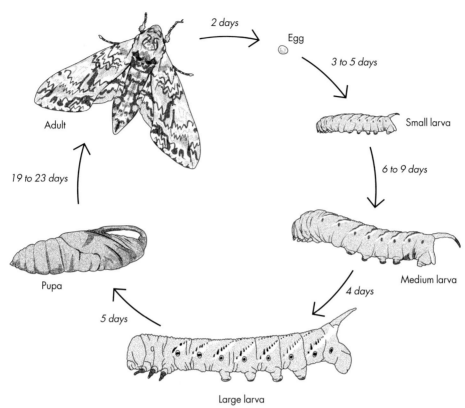

RIGHT | Life cycle of Tobacco Hornworm (*Manduca sexta*). Development times for all stages and instars are temperature dependent.

All moths and butterflies have four life stages: the egg, larva, pupa, and adult. While moth guides focus on the last of these, because the adult stage is the most conspicuous in habit, the early stages occupy most of a moth's life and have greater ecological and economic importance.

EGG

The smallest Geometridae (*Idaea*) with a wingspan as modest as 0.24 in (6–7 mm) may lay fewer than a dozen eggs. Giant ghost moths (Hepialidae) of Australia can produce more than 40,000 eggs, opting for a sweepstake strategy in which the female simply broadcasts eggs as she flies around seemingly suitable habitat. Something in the range of 40 to 200 eggs per female is typical. Moth eggs can be flat, discoidal, round, or spindle-like. Most are white to green in color, but many microlepidopterans have transparent eggs scarcely detectable to the human eye. A few are orange or red, many are black, and some degree of patterning occurs in many lineages. An egg's outer surface (the chorion) may be smooth or wonderfully ornate. Among the latter are those that can trap air in the interstices of the chorion when submerged, and

LEFT | Flightless female of Vapourer Moth (*Orgyia antiqua*) surrounded by her eggs.

BELOW | Shapeshifters. Moth caterpillars may change dramatically in form and color through their development, with most changes occurring during molts. The early (left) and last (right) instars of the Funerary Dagger (*Acronicta funeralis*) are scarcely alignable.

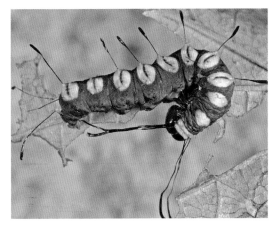

in so doing allow the egg to "breathe" for months underwater. Such is commonplace for riparian and wetland moths, and represents a novel strategy to protect the eggs from desiccation, harsh weather, and the normal sweep of natural enemies that prey on eggs. Upon hatching, many moth caterpillars consume the chorion.

LARVA

While most moths are only known from their adult stage, it is the larval stage that is longer lived, that moves more nutrients to other organisms on the tree of life, causes more damage to our gardens, crops, and forests, and is more crucial for the success of conservation efforts aimed at rescuing a declining species. Caterpillars go through several instars as they grow and mature. Most species have five instars, but some have only three or four, whereas large species may pass through six or seven, with the number of instars sometimes varying by diet and sex (females sometimes have an extra instar). While the norm is for early, middle, and late instars to resemble each other, some caterpillars are shapeshifters that change dramatically in shape, color, and behavior over the course of their development. In the extreme, larval development is hypermetamorphic with the different larval forms specialized for entirely different lifestyles.

A caterpillar has a singular mission: to eat and not get eaten. Some will increase their mass 2,000-fold or more from first to final instar.

LEFT AND BELOW | A variety of moth pupae. (*Left*) Opened cocoon of Indian Moon Moth (*Actias selene*). (*Below left*) Pendant cocoon of a Urodidae. (*Below*) Male bagworm moth (*Cryptothelea gloverii*) eclosing from both its case and pupal shell.

Species inhabiting warm, ephemeral conditions, including many desert-dwelling taxa, can finish their larval development in as little as two weeks and complete their life cycle—egg to adult—in three weeks. Some species of flower moths (Noctuidae: Heliothinae), which feed on the nutritious seeds of their host, complete their development in as little as 19 days.

The fully fed caterpillar, the prepupa, is responsible for seeking out a pupation site and fashioning a cocoon or pupation cell. Feeding ceases and the gut is evacuated. Dramatic changes in coloration may occur at this time—typically colors fade and the integument takes on a subtle shiny aspect; some turn bright red, especially among microlepidopterans. The prepupal stage can last from a few hours to many months in moths that overwinter in this stage; in prodoxids (page 99) the stage may last for decades.

PUPA

Between the caterpillar and adult stages is the pupal stage. While seemingly inactive and relatively immobile and defenseless, this is a stage of tremendous change and reorganization. The muscles and organs of the caterpillar are digested and reshaped into those of adults for flight and reproduction, a process that can take as little as a few days, but more commonly two to three weeks. In nearly all moths, pupation takes place in a sheltered site in a cell below ground, in a leaf shelter, or a silken cocoon spun by the prepupa.

A few, such as plume moths (Pterophoridae) and sterrhine geometrids, have fully exposed pupae (chrysalises), as occurs in most butterflies.

Moth pupae tend to be rather unmodified, bullet-shaped creatures. Most are a nearly uniform shade of brown, but taxa with chemically protected pupae may be brightly or otherwise boldly colored—these are often made visible in wispy or open-mesh cocoons. The precursors of adult appendages—antennae, mouthparts, wings, and legs—are fused to the pupa casing in all but the most ancient moth lineages. In most pupae the body segments are fused except for a few abdominal segments that allow the pupa to twirl, contract, and extend. Microlepidopteran pupae are often armed with a battery of backward-directed teeth over the dorsum of the abdominal segments; these teeth help push the pupa forward, through the cocoon wall or out of the pupal cell prior to eclosion (emergence from the pupal shell). The vertex of the pupa sometimes is drawn out into a point, which additionally helps the pupa free itself from the cocoon.

BROODS

At high latitudes and elevations there is often just a single generation per year, with some species taking two years or more to develop. A record holder for the most extended life cycle is the Arctic Woolly Bear, *Gynaephora groenlandica*, which takes up to seven (or more) years to mature. Activity of boreal and temperate moths tends to be regulated by favorable temperatures and day lengths. Moving to the equator, many moths add generations and, where freezes are rare, can

BELOW | Arctic Woolly Bear (*Gynaephora groenlandica*). Larval development is slow in this tussock caterpillar of high latitudes of the northern hemisphere. Individuals commonly take several to many years to mature.

LEFT | Bogus yucca moths of the genus *Prodoxus* are diapause record holders, capable of holding over in their larval tunnels as prepupae in the floral stalks of yucca for as many as 30 years.

become year-round breeders if appropriate growth is present. In drylands and other seasonal ecosystems both adult and larval activity are often highly synchronized to periods of favorable moisture and plant growth rather than day lengths. For example, moths that are strictly active during the summer monsoon in the Sonoran Desert of North America are often multibrooded and active from spring into November in the neighboring Chihuahuan Desert of western Texas. Facultative broods, where just a fraction of a population emerges under a given set of weather conditions, add further possibilities, especially where climates are variable.

DIAPAUSE

Moths that live in seasonally harsh environments have metabolically inactive diapausing stages that may last from a few weeks to years, allowing them to endure extended periods of cold and dryness, such as winters in temperate, boreal, and polar ecosystems and the dry seasons in deserts and grasslands. The period of stasis is under hormonal control and generally programmed to precede the period of inclement weather. In most insects, the period of diapause is usually triggered by changing day lengths, which allows the insect to prepare metabolically prior to the onset of bad weather or the deteriorating quality of their food resources.

Moths may overwinter in any of the four life stages. A large percentage of species overwinter as eggs; some, especially among the ground feeders, as caterpillars; the majority of moths as pupae; and a small fraction as adults. In the last case, this may mean that the adult will freeze and thaw many times over the winter months. Remarkably, such winter moths prepare themselves in an analogous way as people prepare their cars: we add glycol (aka antifreeze) to our radiators, while insects manufacture glycols and specialized proteins that prevent ice from forming in their cells. Moths dwelling in deserts and drylands mostly diapause belowground or are seasonal migrants that move between areas of favorable conditions. The longest diapausing moths invariably dwell in deserts, where rainfall is scarce and unpredictable. So far, the record holder is a bogus yucca moth in the genus *Prodoxus*: a single larval cohort, secreted in their larval tunnels as prepupae, yielded adults over a period of 30 years.

MOTH SENSES AND SOUND

Moths and other insects have the same five primary senses as humans: vision, smell, sound, taste, and touch. The treatment below is but an introduction to those most relevant to the content of this guide.

VISION

Moth compound eyes are like those of other insects that can detect movement well, but visual acuity is thought to be modest. Relative to our own, their color vision is shifted into high-energy wavelengths, with only ultraviolet (UV), blue, and green sensitivity. Most moths lack receptors for red and infrared wavelengths. Such is generally true for nocturnal insects, and this is why moth collectors, photographers, and watchers deploy lights high in ultraviolet wavelengths to attract moths, and red lights to observe moth activities at night. While moths, being active primarily in the dark, may have less to gain from color vision, they still need to detect and respond to visual cues in order to locate nectar resources and water. So far as known, hawk or sphingid moths, which are important pollinators by night, have the most color-sensitive eyes, allowing them to locate flowers even below tropical forest canopies at night.

Additionally, many adult moths have two ocelli above the antennae that are non-image-forming light detectors. They are thought to play important roles in entraining the timing of daily behavior patterns. Ocelli also serve roles in maintaining the body's vertical orientation, especially while in flight.

Most caterpillars have six image-forming stemmata on each side of the head. These generally have the same set of color-sensitive

BELOW | Moths' eyes range from comparatively large, occupying much of the head (pictured), to small in nocturnal taxa whose primary sensory modality is smell.

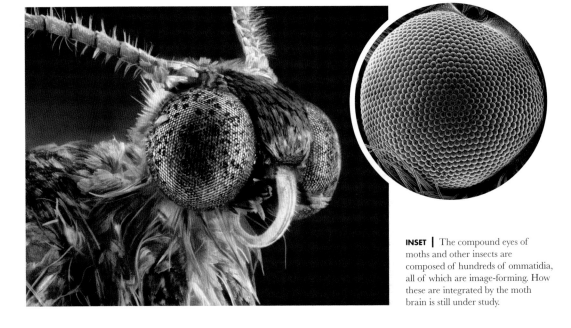

INSET | The compound eyes of moths and other insects are composed of hundreds of ommatidia, all of which are image-forming. How these are integrated by the moth brain is still under study.

ABOVE | The caterpillars of Abbott's Sphinx (*Sphecodina abbottii*) squeak (and bite) if handled roughly.

LEFT | The African Death's-head Sphinx (*Acherontia atropos*)—note the scale pattern of the thorax—produces an audible squeak by forcing air out of its haustellum.

opsins as the adults—that is, they see UV, blue, and green colors. Entomologists have recently learned that caterpillars can "see" with dermal cells. While no images are formed, the cells allow a caterpillar to sense its immediate light environment, and perhaps choose an advantageous resting site. At molts, both caterpillars and pupae of some taxa (for example, ennomine geometrids and swallowtail butterflies) have the uncanny ability to take on a green or brown form that is more cryptic given their surroundings, with the response seemingly linked to feedback from the dermal cells.

SOUND PRODUCTION

Relatively few moths and even fewer caterpillars are known to produce audible sounds. Most commonly, there are moths that squeak or stridulate when threatened, presumably in an attempt to startle their would-be enemy. The famous African Death's-head Sphinx (*Acherontia atropos*) "chirps" by inhaling and exhaling air so quickly that rapid actions cause part of the moth's haustellum to vibrate. Several sphinx moth caterpillars emit sounds. Members of the genus *Amorpha* hiss when alarmed, by forcing air out of their spiracles. In such cases, the insects are harmless and palatable, and as such make for vulnerable prey should their attempts to startle their enemies be ignored.

While audible sound production is rare, many moths use ultrasound—frequencies above those detectable by humans—in courtship. Males of many crambid and pyralid moths "sing" to attract females or advance their courtship attempts by vibrating their wings. Hollow pockets, either opened or closed on the wings of moths, presumably serve as resonating chambers.

Ultrasound is also used in defense, either to jam the echolocation frequencies used by hunting bats or to advertise a moth's unpalatability. The latter is widespread among tiger moths, which are

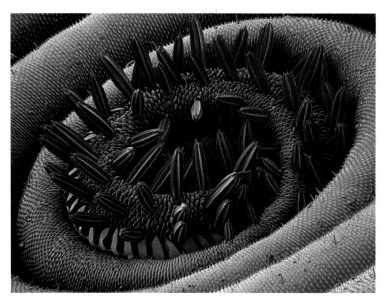

LEFT | Colorized SEM of a moth haustellum: the tongue (light blue) is coiled up; the red setae, concentrated distally on the haustellum, are chemoreceptors that taste the fluids before these are drawn into the mouth.

renowned for their ability to sequester toxic alkaloids from their hosts as well as manufacture histamines and other bioactive defensive compounds. In tiger moths, the ultrasound is created by the rapid flexing of a corrugated section of their integument (the tymbal), below the base of the hindwing on the side of the thorax. Some male and female sphinx moths are able to produce ultrasonic jamming frequencies with anatomically different genitalic structures!

SOUND RECEPTION

Essentially all macrolepidoptera that are nocturnal have some means of avoiding or otherwise thwarting bats, which, to a large measure, are moth hunters. Approximately 70 percent of the world's 1,400 species of bats are insectivores. Moths have evolved ears or some functionally equivalent ultrasound detector independently many times since bats started hunting night skies for insect prey. Owlet moths (Noctuidae, Erebidae, and kin) have a sophisticated ear, specialized for detecting the echolocation frequencies used by bats, just under the hindwing. Geometrids and pyraloids independently evolved ears at the front of their abdomen that sit in the sinus where the thorax and abdomen join. Those of hedylids are on the wings. Of note are the ultrasound detectors of sphinx moths, located at the apex of the labial palpi coupled with cells of the labral pilifers.

TASTE

Moths, like other insects, have four principal tasting organs: the antennae, mouthparts, tarsi, and ovipositor. Moths have the same principal taste senses as people—that is, they can differentiate sweet, sour, bitter, and acid compounds. The most common type of taste receptor is a cone-like sensillum that has apical pores that allow fluids to be drawn into the cells. Within a given sensillum there might be one to four taste cells that respond to different molecules. In adult moths much of the initial tasting is done with the tarsi when the adult alights on, or walks over, a substrate. If a desired resource is detected, the haustellum, with a unique bank of sensilla, is then uncoiled and dipped into the resource.

Antennae, while capable of tasting, are not used to taste foods in moths; instead, the antennae are specialized for the detection of volatile (airborne) molecules. The taste receptors at the terminus of the female ovipositor play a crucial role in the biology of moths. The ovipositor lobes, situated at the ends of long, chitinous rods (the posterior apophyses), bear numerous chemosensory setae that assist the female in egg placement, as the eggs are being passed from the abdomen. The ovipositor lobes are rich in sensilla and gustatory cells specialized for the detection of secondary plant compounds unique to the moth's host plant, and even provide the female with information of the nutritional appropriateness of the ovipositional site.

A caterpillar's decisions about what to ingest are critical, especially given that many of the secondary plant compounds manufactured by plants are toxic, intended to dissuade herbivory (see page 43). An ill-made decision can be fatal. Caterpillars have a rich assortment of pre- and postoral sensilla that assess the chemical nature of their food. The caterpillar's short antennae bear important preoral taste receptors. The most essential sensilla relevant to ingestion are located on the maxilla and labium; other sensilla may occur on the mandible and labrum. The sensilla typically house three or four taste cells, each of which is specialized to assess a single class of molecules. For example, the labial or maxillary palp often has one of these sensilla dedicated to the detection of compounds uniquely possessed by the species' preferred host plants.

BELOW | The antennae of giant silk moths are greatly enlarged, especially in males.

INSET | Colored SEM showing an abundance of chemosensory setae specialized for detection of the female sex pheromone (in males) and host plants (in females).

LEFT | Fiery Clearwing (*Pyropteron chrysidiformis*). Synthetic chemical lures mimicking the female sex pheromone of many sesiids can be purchased to attract males.

SMELL AND PHEROMONES

Because moths are mostly nocturnal and visual signals are rendered ineffective, airborne chemicals (scents) have come to play especially important roles in the lives of moths for nectar resources, courtship and mating, and host-plant detection by ovipositing females. The hallmark antennae of male moths are another testament of the importance of olfaction to the biology of moths. Moths are renowned for their ability to detect minute chemical concentrations—the many-branched antennae of some male moths rank among nature's most sensitive chemical-detection systems. Much of the antenna's elaboration and complexity is focused on the detection of the sex pheromones released by calling females, which are typically emitted from a special abdominal gland upon emergence. It is believed that the males of some giant silk moths can detect calling females at a distance of a half mile or more, when air movements are slow and unidirectional.

Male pheromones are chemically highly diverse, and, in contrast to those of females, many are detectable to humans, with a spicy, pleasant odor. In Noctuoidea and many other lineages, the pheromones are manufactured in the pupal stage, such that once a male ecloses his allotment is finite. Because most male pheromones tend to be volatile and thus vulnerable to evaporation, the androconia (scent-disseminating scales) are often tucked in a pocket or folds and are only deployed during courtship displays. Androconia are most often present when closely related moths share a common host plant, such that the probability of an errant coupling with the wrong species is elevated.

SYNTHETIC PHEROMONES

Many synthetic analogs of female sex pheromones are now commercially available, especially for common pest species. Scientists can use synthetic attractants to monitor the presence and abundance of pest species so as to better time pesticide applications. They aid in the early detection of exotic or introduced species. And, deployed in high density, they serve to trap-out the males of local populations. This last insect-control method is often attempted to prevent the establishment of a new, spatially localized pest.

Moth collectors and photographers can also purchase synthetic pheromone lures to attract species of special interest. Indeed, synthetic lures are practically the only way to see some of the most beautiful and unusual moths treated in this volume, such as the clearwing moths (Sesiidae).

ECONOMIC IMPORTANCE OF MOTHS

Moths have many positive and negative consequences for humanity, and for nature. The value of insects to maintaining the planet's terrestrial ecosystems has never been estimated but is undeniably priceless. As noted above, caterpillars and moths are among the most diverse, numerically abundant animals, and, most importantly, account for much of the biomass available to nature's insectivores, and then by extension all animals that eat insectivores. In this section, we skip past the many ecosystem functions of Lepidoptera, and focus on their immediate importance to humanity and our economies.

SILK

Caterpillars are renowned for their capacity to produce silk—a proteinaceous fiber valued for its satiny feel, shimmering appearance, tensile strength, hypoallergenic properties, and light weight. While the ability to produce silk is almost universal among caterpillars, especially in the last, cocoon-spinning instar, virtually all the silk used

RIGHT | Cocoons of Domesticated Silk Moth (*Bombyx mori*)—when unraveled, the single fiber secreted by the prepupal caterpillar may extend for more than a half mile.

BELOW | Domesticated Silk Moth caterpillars feeding on mulberry leaves.

in commercial production comes from the Domesticated Silk Moth (*Bombyx mori*), which has been entirely domesticated for the production of its silk. Its cocoon is immersed in boiling water, then unraveled to make the highest grade of silk, mulberry silk, used in textiles that include kimonos, saris, pajamas, lingerie, shirts, ties, sheets, and haute-couture clothing. Silk is seeing increased use in medicine—for example, as scaffolding for repairing burns and other wounds as well as the repair and healing of bone, cartilage, tendon, and ligament tissues. China and India are the world's largest producers of silk. Global markets are presently valued at about $20 billion annually, and are expected to rise beyond $30–$35 billion by 2030.

MOTHS AS PESTS

Caterpillars are significant insect pests, causing billions in annual damage to forests, orchard and field crops, and once siloed, to grains that are spoiled by a number of pyralid stored product pests. Introduced populations of Spongy Moth (*Lymantria dispar*) have killed millions of oak trees throughout the northeastern United States at tremendous economic and ecological costs. Several dozen noctuids are chronic pests of field

BELOW | Moths include many important forest and crop pests. The Corn Earworm (*Helicoverpa zea*) is estimated to be responsible for crop losses exceeding US$1 billion annually.

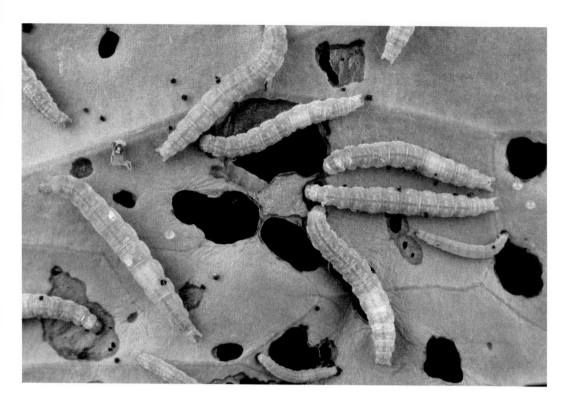

ABOVE | The Diamondback Moth (*Plutella xylostella*) is a global pest of crucifers (Brassicaceae), consuming all aboveground tissues and soiling the crops with their droppings.

crops and responsible for much of the planet's pesticide use, which, in addition to killing caterpillars, elevates human cancer risks, underlies thousands of accidental poisonings—some of which are fatal—and accounts for untold nontarget effects in croplands and adjacent natural areas. Among the most infamous pests are the Cotton Bollworm (*Helicoverpa armigera*), Diamondback Moth (*Plutella xylostella*), Taro Caterpillar (*Spodoptera litura*), and Fall Armyworm (*Spodoptera frugiperda*). Several noctuids, in particular, commonly reach densities where entire fields are laid waste, accounting for much famine and hardship in areas of the world where pesticides are not readily available, too expensive, or the farmers lack the capacity to apply the pesticides in a timely fashion or at an appropriate scale. Caterpillars of the diminutive Diamondback Moth are chronic pests of crucifers worldwide—broccoli, Brussels sprouts, cabbage, cauliflower, and kale—and their ravages exceed $4 billion annually to the world economy.

Grains, birdseed, cereals, flour, dog food, nuts, dehydrated fruits, and other dried foods are attacked in storage by phycitine pyralids and other lepidopterans. The larvae of the Mediterranean Flour Moth (*Ephestia kuehniella*) and Indianmeal Moth (*Plodia interpunctella*) are worldwide pests, soiling stores with their silk and feculae. In addition to loss of product due to contamination, these and other stored product pests can trigger costly product recalls, consumer complaints and litigation, and damage brand reputation.

Several species of clothes moths (Tineidae) are chronic pests of woolens, furs, and feathered

LEFT | Case-bearing Clothes Moth (*Tinea pellionella*) caterpillar extended from its silken case, into which it has incorporated some red fibers from the textile it's consuming. Woolens and other fabrics made of animal hair are targeted by some tineids.

items worldwide. Their casemaking caterpillars have the exceptional ability to digest keratin—the chemically resistant protein that makes up claws, fur, hair, hooves, nails, and horns of mammals. Clothes moth larvae are most problematic with stored fabrics, clothing, wall hangings, and tapestries that are not periodically washed.

CATERPILLARS AND MOTHS AS HUMAN FOOD

Caterpillars are highly nutritious—they are rich in protein, including all essential amino acids, unsaturated fatty acids, minerals, and vitamins. When dried, their protein proportion is comparable to that of raw beef and is far more sustainable environmentally, requiring less land, less water, and yielding minimal greenhouse emissions. Caterpillars, and less frequently moths, are important food sources for many indigenous peoples, especially in Asia, Africa, and Latin America. Large caterpillars, in particular those of giant silk moths (Saturniidae) that reach high local population levels, are frequently a target. Most famously, the Mopane Worm (*Gonimbrasia belina*) is eaten across southern Africa, either dried or cooked in prepared dishes. In Veracruz, Mexico, the *Arsenura* caterpillar, another giant silk moth, is targeted for human consumption. Australian tribes consume caterpillars, especially large-bodied, wood-boring cossids and hepialids. Among the most common is the Witchetty Grub (*Endoxyla leucomochla*), a cossid that bores in the stem and upper roots of various *Acacia*.

Indigenous groups in Mexico collect the Maguey Worm (*Comadia redtenbacheri*), a cossid that bores into agave leaves and roots. Single plants may yield hundreds of caterpillars. The larvae, known as *chilocuiles*, *chinicuiles*, or *tecoles* in Mexico, get increasingly red as they mature—prepupae are bright red. As such, the caterpillar also has become widely known as the red Maguey Worm and *gusano rojo*. Notably, Maguey Worm is the caterpillar most commonly found swirling at the bottom of some mezcal bottles.

The boiled pupae of silk worms, a by-product of sericulture, are an enormous nutritional resource. China alone produces more than 100,000 tons of Domesticated Silk Moth pupae annually. While the pupae are canned and sold as human food in Asia, more commonly they are dried and used as an additive in human foods or processed into animal feeds, especially for chickens.

In Bhutan and Tibet, the caterpillars of several species of *Pharmacis* and *Thitarodes* (Hepialidae) are highly valued as an aphrodisiac, a remedy for cancer and a sweep of other ailments, and a prophylactic for still other maladies, but only after the larvae have been attacked and mummified by a fungus (*Ophiocordyceps sinensis*)! As far as I am aware, there is scant evidence that the fungus zombies cure any ills, beyond what an alternative placebo would treat. The dried caterpillar–fungus cadavers are worth twice their weight in gold. Well-preserved specimens may sell for $140,000 a kilo. In Bhutan

LEFT | Large caterpillars such as this African saturniid are commonly roasted and sold in Asian, African, and Latin American markets.

BELOW | In Central and South America, caterpillars such as the Maguey Worm (*Comadia redtenbacheri*) serve as a protein supplement.

LEFT | Roasted mopane caterpillars (*Gonimbrasia belina*) for sale in a market. Late instars of this large silk moth are an important source of protein for many peoples across southwestern Africa.

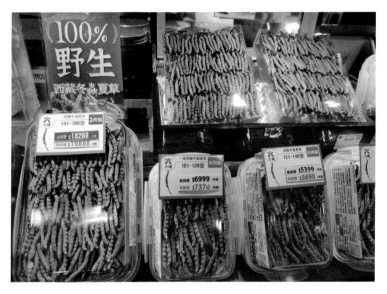

LEFT | Ghost moth (Hepialidae) "fungus caterpillars" for sale on a Chinese market. The mycelia-ridden caterpillars are ground up and used as traditional folk medicine to treat various maladies and as an aphrodisiac.

the harvesting and sale of the mummies is a significant source of income for many rural families, with school activities sometimes scheduled around the annual appearance of the fungal fruiting bodies.

Adult moths are much less often consumed, with one important exception: Australia's Bogong Moth (*Agrotis infusa*), which gregariously aestivates in caves and scree high in the mountains of southeastern Australia. Each summer billions of the adults would migrate to the caves and other sheltered sites to await fall rains and the greening of the larval feeding sites. For centuries, Aboriginal tribes would come to the Bogong's montane aggregations to feast on the moths for weeks at a time. Changes in land use, agricultural intensification, and a severe drought in 2017 so diminished the moth's numbers that today many of the caves are empty, and the species is now regarded as endangered (see also page 55).

MOTHS AND PLANTS

Insects and plants have been evolving together for at least 400 million years. The fates of the two groups are inextricably intertwined and have spawned countless interactions over geological time. The three most archaic moth lineages, the Micropterigidae, Agathiphagidae, and Heterobathmiidae, have retained host associations similar to those believed to have existed at the time of their origin. In fact, there is much reason to believe many modern-day host associations are reflective of those of their ancient ancestors.

The vast majority of today's estimated 300,000 plus species of Lepidoptera feed on angiosperms (flowering plants). Angiosperms started diversifying as early as the Jurassic, or even before, at a time when cone-bearing plants (gymnosperms), ferns, club mosses, liverworts, and mosses were the ecologically dominant plants across Earth's terrestrial ecosystems. However, that all changed some 125 mya when angiosperms radiated and catapulted themselves (with the help of insects as pollinators) into ecological dominance across the planet. Moth–plant associations span the gamut from highly mutualistic and codependent to highly antagonistic, with caterpillars sometimes fully consuming their host plant, or in the case of shrubs and trees, defoliating, and not infrequently killing their hosts.

SECONDARY PLANT COMPOUNDS

Plants defend themselves from herbivores and other enemies using a battery of physical defenses that include spines, hooks, dense hairs, and

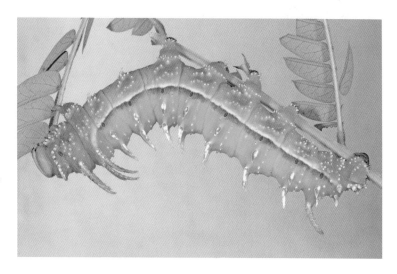

ABOVE RIGHT | A fairy moth (Adelidae) visiting a flower. The effectiveness of adelids and most other archaic moths as pollinators is in much need of study. Their cousins, the Prodoxidae, are renowned for their pollination services, upon which yuccas are entirely dependent.

RIGHT | A *Syssphinx raspa* caterpillar. While eating different plants across their range, local populations of many moths specialize on just one species at a given place—here, prairie acacia in southeast Arizona.

STRUCTURE OF RETRONECINE

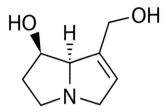

ABOVE | Retronecine, a common pyrrolizidine alkaloid (PA), is a powerful secondary plant compound, toxic to liver tissues.

hardened tissues as well as a galaxy of chemical defenses, or secondary plant compounds (primary compounds are those needed for photosynthesis, respiration, and other basic metabolic functions). Secondary plant compounds known to deter herbivory by caterpillars include alkaloids, nonprotein amino acids, cyanide-producing compounds, cardenolides, glycosides, phenolics, terpenes, salts, silica, several classes of sulfur-containing compounds, and still others.

Any moth lineage that evolves the capacity to avoid, detoxify, tolerate, or sequester the secondary metabolites of a diverse, ecologically abundant plant lineage has much to gain evolutionarily. Should the moth lineage radiate in turn, it poses increasing selective pressure on the plant group. And as the herbivory pressures increase on a plant lineage, there would be increasing fitness benefits to any member of the plant lineage to evolve some *new* secondary compound or other means of self-protection. Upon doing so, the plant taxon, now freed of much herbivory, would be poised to radiate in turn.

Such a situation would represent but two bouts in an ancient arms race that has been ongoing between insects and their host plants for millions of years dating back to at least the Permian Era—an ongoing coevolutionary battle that may account for much of the planet's species diversity.

Pyrrolizidine alkaloids (PAs), common to many borages (Boraginaceae), composites (Asteraceae), legumes (Fabaceae), and orchids (Orchidaceae), underlie much of the evolutionary successes of several groups of moths, and in particular the tiger moths (Erebidae: Arctiinae). PAs are potent liver toxins and mutagens when consumed by vertebrates. Tiger moth caterpillars that sequester PAs as caterpillars render themselves unpalatable to lizards, birds, and bats, and warn of their toxicity with yellow, orange, white, and black colors, in both the larval and adult stages, and in some cases, even the egg and pupal stages.

LARVAL DIETS AND HOST-PLANT SPECIALIZATION

Perhaps 98 percent of all Lepidoptera are herbivorous, with some 85 percent of these being host-plant specialists as caterpillars that eat just one or a few related plant species. The remaining taxa are either oligophagous, feeding on plants in just two to three plant families, or polyphagous, feeding on plants in four or more families. The ecological and evolutionary advantages and disadvantages of these strategies are active areas of inquiry among ecologists. Surely part of the explanation is that specialists become adept at detoxifying or otherwise processing the potpourri of secondary plant compounds that plants manufacture to protect themselves from herbivores and other enemies.

There is another eco-evolutionary advantage to host-plant specialization, in that, over time, the caterpillar of a specialist species can be shaped by natural selection to resemble its host plant—in color, shape, texture, reflectance—and adopt appropriate behaviors, to lower its apparency to visual predators such as lizards, monkeys, and

ABOVE LEFT | A male approaching a female Bella Moth (*Utetheisa ornatrix*). Males transfer PAs to their female partners during mating.

ABOVE | A Bella Moth larva sequesters PAs from its host plant. All four life stages of Bella Moths may be protected by PAs.

especially birds. This argument becomes more compelling when considering the fate of host-plant generalists, which are on different plants from generation to generation (and even day to day), and as a consequence are not well matched to any single plant.

The diversity of moth species hosted by a plant is largely a function of the host plant's geographic range, architectural complexity, and range-wide abundance; other determinants include the plant's apparency, physical and chemical properties, and degree of taxonomic isolation (for example, the number of congeners or family members growing nearby).

In addition to plants, there are moth caterpillars that feed on fallen leaves and flowers, fungi, lichens, and algae. Tineidae are unique among lepidopterans in that they can digest keratin, which allows their caterpillars to mature on diets of antlers and horns, feathers, fur (including wool), owl pellets, and turtle shell.

PLANT-FEEDING GUILDS

While most caterpillars feed on leaves, essentially all plant tissues are consumed by Lepidoptera. Many lineages, including Noctuidae, Geometridae, Pyralidae and Crambidae, Plutellidae,

ABOVE | Not all caterpillars eat green plants. Footmen or lithosiines radiated on lichens. The long setae on this *Eudesmia arida* caterpillar keep many of its smaller enemies at bay.

LEFT | A few small lineages of owlet moths feed principally on fungal hyphae. So far as known, all members of *Metalectra* are dietary specialists on fungi (and can be raised on store-bought mushrooms). This example is *Metalectra diabolica*.

Coleophoridae, Adelidae, and Agathiphagidae, among others, target fruits and seeds. Borers in nonwoody stems include many Noctuidae, Tortricidae, and Momphidae. These may feed on the entirety of the stem, while others specialize on certain tissues such as the meristems, pith, epidermis, or the outer photosynthetic tissues. Most gall-forming Lepidoptera are stem gallers.

Root feeding is surely more widespread than appreciated. Borers that enter above or just below the soil and tunnel into the roots include Hepialidae, Noctuidae (especially noctuines), and Sesiidae. Lepidoptera that can burrow through sand and friable soils—such as Hepialidae, Noctuinae, and Crambidae—will feed externally on roots.

Leaf-mining has been reported in some 20 families of Lepidoptera, and is especially common among archaic families and smaller microlepidopterans. Given that the leaf is also the caterpillar's environment, the host ranges of leafminers tend to be more specialized. Species-rich families include the Bucculatricidae, Gracillariidae, Nepticulidae, and Tischeriidae.

POLLINATION

Moths are among the most important insect pollinators, second only to bees. But just how important moths are to flowers remains one of the more poorly documented aspects of their biology because so much of their activity occurs after nightfall. Moth-pollinated flowers tend to be white, fragrant, and make their nectar available at night. Many members of the evening primrose (Onagraceae) and four o'clock (Nyctaginaceae) families do not even open their flowers until late afternoon or dusk.

Nearly all "moth flowers" produce sweet scents. Indeed, some of the flowers with the most pleasant fragrances are moth pollinated. The wonderful aromas of carnations, gardenias, honeysuckle, lilacs, narcissus, and the queen of all, jasmine, are volatiles, manufactured by plants to encourage pollen transfer by moths. Common floral scents include benzaldehyde, lilac aldehydes, linalool, methyl benzoate, and phenylacetaldehyde.

BELOW | Fruits and seeds like corn are targeted by many moth caterpillars, here a Western Bean Cutworm (*Richia albicosta*).

BOTTOM | Serpentine mines of *Phyllocnistis populiella* on aspen. The mature larva, visible on the lower left leaf, is beginning to spin its pupal crypt—a minute pinched fold fashioned along the leaf edge.

ABOVE | *Hemaris* are small day-flying sphingids that are often mistaken for bumblebees or hummingbirds. About 20 or so species are widely distributed across the northern hemisphere.

NURSERY POLLINATION

Of special note are nursery pollination systems wherein a moth species is both a dedicated pollinator and consumer of the host plant's seeds and fruits. Upon transferal of pollen to an appropriate flower, the female then lays her egg(s) in or on the flower—her offspring will mature on the developing seeds and associated tissues. The most renowned example of a nursery pollination system occurs in yuccas and yucca moths—the two lineages are engaged in an obligatory mutualism where the survival of each is interdependent and absolute. A more globally widespread and understudied nursery pollination system is that of *Silene* (family Caryophyllaceae) and the coronet moths (*Hadena*; Noctuidae) that see to the pollination of the world's 900 species of the genus.

Sphinx moths are recognized as important pollinators, especially in arid regions and the tropics. They are strong fliers, have the longest tongues of any insects, and at least some are thought to follow traplines—that is, they can remember the locations of widely scattered and sometimes distant nectar sources—an attribute that can be especially important in the tropics where many trees grow in low densities, often well removed from one another. Moreover, sphingids have acute color vision—the most sensitive known across the animal kingdom—enabling them to detect colors under extremely low light conditions and ensuring that they will be able to navigate to flowers even on new moons and under closed forest canopies.

Madagascar's Comet Orchid (*Angraecum sesquipedale*) has a spectacular white, fragrant flower with an enormous nectar spur that often exceeds 10 in (25 cm) in length. As its name implies, the spur is a floral extension where a flower produces, and more to the point, caches its nectar reward. Charles Darwin, upon examining a bloom of the Comet Orchid, predicted that there must be a moth with a tongue long enough to reach to the bottom of the orchid's spur. It was not until 21 years after his death, that the moth *Xanthopan praedicta* was discovered—a moth known today as Wallace's Sphinx Moth—whose lingual siphon may exceed 11 in (28 cm) in length.

MOTH BEHAVIOR AND ECOLOGY

Just a few areas of the extraordinary palette of behaviors and ecological interactions of moths can be shared here. Because so much of what a moth is and does happens at night, suffice it to say, much more remains to be written. To some degree, the same is true for their caterpillars.

COURTSHIP AND MATING

Much regarding the reproductive behavior of moths appears elsewhere in these introductory pages and in the taxon profiles that follow, in large measure because moths, the ultimate stage of these insects, have a principal mission: to propagate. Those moths with a short adult lifespan are especially singularly minded: upon eclosion the female must find a mate, court, pair, and then get to the business of dispersal and oviposition. For males, only the first three of these activities will occupy their psyche. The major exceptions to the above are those species that enter a reproductive diapause before mating—a behavior common among those moths that hibernate or aestivate as adults.

In nearly all moths, the females produce the principal sex pheromone—that is, the scent to which males will orient and compete for mating opportunities. In most species, the primary sex pheromone is released by females from specialized abdominal glands soon after she has eclosed, typically immediately after her wings have fully expanded and before she has taken her first flight. In exceptional cases, such as ghost moths, this signaling system may be reversed.

Moths couple end-to-end. In Saturniidae, the male and female frequently remain in copula until nightfall of the next day. In Ditrysia, which account for about 98 percent of extant moth diversity, females have a separate reproductive opening solely for copulation. Further anterior in the female's abdomen, Monotrysia and Ditrysia have a sack-like enlargement, the bursa copulatrix that receives the male spermatophore, and less commonly multiple spermatophores. From the bursa the sperm must swim through a narrow duct, past the common oviduct, and make their way into a special gland, the spermatheca, where they await the passing of unfertilized eggs.

The male spermatophore and the bursa that can house it may well represent key innovations for moths. Spermatophores, depending on the taxon, routinely contain proteins, lipids and

LEFT | Fall-generation adults of Herald Moth (*Scoliopteryx libatrix*) (Erebidae) enter a reproductive diapause, seek out caves in which to overwinter, then mate and lay their eggs in the spring. An aggregation pheromone may be involved as the hibernating moths are often found clustered.

LEFT | A recently emerged Buck Moth (*Hemileuca maia*) "calling"; that is, emitting the female sex pheromone from a gland near the terminus of her abdomen.

sterols, carbohydrates, minerals, vitamins, and defensive chemicals, in addition to sperm, that will contribute to the fitness of both the female and the couple's offspring. Stated differently, the bursa allows the male to make massive energetic and defensive contributions to his partner and offspring that have played important roles in the diversification and ecological successes of Lepidoptera.

The extended duration of mating likely relates to the time necessary for the male to transfer his spermatophore to the female. In some species the male will lose considerable mass over the course of a single mating. There is some evidence that virgin males make preferred partners for females, and that, once-mated, males may have long refractory periods before they are able to constitute a second spermatophore, and even then, it may be of lesser value.

TERMINALIA AND SEXUAL CONFLICT

The male and female terminalia of moths tend to be quite complex morphologically, with the two fitting together like a lock and key. The male genitalia are particularly elaborate. Whether this complexity is due to sexual selection and the ancient battles of the sexes fighting to control paternity, or the need to prevent mating mistakes with closely related species, is a fascinating subject. But for whichever reason, the genitalia and associated secondary sexual structures are among the most rapidly evolving anatomical features of moths. As such, they are important for making species-level identifications by moth taxonomists, collectors, extension entomologists, and others.

In moths, as in many other animals, there is ongoing conflict between the sexes, with each having evolved multiple measures to control facets of reproductive interactions. At the point of courtship, the female has much control in that she determines when to call, when to proceed past courtship to accept a male partner, and when to expose her abdomen to her suitor. A female disinterested in a pursuant male can stop calling, fly off, or raise her abdomen to make it unavailable. Additional controls are built into the female's anatomy. In some moths, the ductus bursae is corkscrewed, which can prevent entry of the aedeagus of a closely related species that has not "coevolved" to meet the biomechanical demands required by the female's anatomy. Sperm precedence plays an important role. After mating, sperm are stored in the spermatheca, atop the female oviduct, where they await the passage of unfertilized eggs out of the common oviduct. The sperm near the entry duct are most likely to fertilize a given egg. Thus, a female that accepts a pairing with a second male will, in so doing, favor parentage by her most recent suitor.

In their attempts to control paternity, males have evolved numerous strategies to discourage subsequent pairings. The male spermatophore,

BELOW | Lateral view of a ditrysian moth abdomen. Note the separate openings for egg deposition (terminus of oviduct) and mating (ostium bursae).

FEMALE REPRODUCTIVE ANATOMY

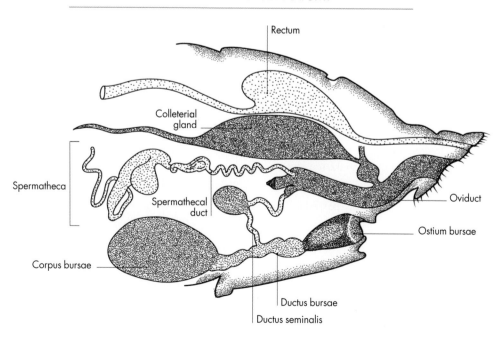

LEFT | In some hepialids the mate-signaling system is reversed: virgin females must fly upwind, tracking a sex pheromone, to locate a calling male. Here, *Phymatopus hectoides* is releasing his pheromone from hundreds of androconia borne on the hind tibiae. (During the day the tibiae and androconia are tucked into deep pockets on the sides of his abdomen.)

transferred to the bursa copulatrix during mating, can be so large in some taxa that the female is incapable of accepting another mate. In other taxa, the female can pair again, but only after she has digested the first spermatophore, which can take days. Males of some moths leave pheromones behind that render a female less attractive, at least through some refractory period. Bordering on the sinister, a number of moths have sharp, deciduous spines on their aedeagus that are left behind in the female genital tract after mating to thwart mating attempts by subsequent males.

PHEROMONES

Moths produce many different sex pheromones, with those of the females being chemically dissimilar to those produced by males. Female sex pheromones tend to be straight-chain, 8- to 20-carbon aliphatic compounds, with one or two double bonds and a chemically active moiety (acid, alcohol, aldehyde, ketone). As a rule, female sex pheromones are odorless to humans.

Because female sex pheromones are rather chemically simple in their diversity, there can be cross-reactivity among both closely and distantly related species. In most cases, and especially when two or more congeners are sympatric (that is, active at the same locale), the exact blend of constituent molecules in the sex pheromone cocktail will differ. Species-specific signaling can also be attained by having pheromone blends with differing chiralities. Cross-reactivity and mating mistakes also can be prevented by related species calling at different times of night, being active during different seasons, or by occupying separate habitats or regions.

Males may also employ pheromones that are important for female acceptance in some lineages. Most male pheromones are released from specialized scales called androconia that are deployed only in the vicinity of the calling female, as part of a pre-mating courtship. The androconia responsible for pheromone release tend to have an elaborate ultrastructure, which provides an exaggerated surface area for rapid volatilization of the pheromonal compounds. Male scent scales and brushes are most commonly found in association with the genital capsule, adjacent abdominal segments, or on the wings, but also occur on the legs (especially the hindlegs, which are those most proximate to the genitalia); less commonly, they occur on antennae and labial palpi.

Androconia can be single specialized scales or, more commonly, scale clusters that form brushes, "hairpencils," scent patches, or, in special cases, incorporated into elaborate courtship organs. A storied example is that of the abdominal courtship brush present in many Noctuidae. The organ, secreted in a pleural fold of the abdomen until needed, consists of two levers, an elaborate distal androconial brush, composed of dozens of golden, pheromone-laden setae, and a pheromone-producing gland. Immediately prior to coupling, the brush is pulled from the pocket and the androconia splayed.

MIGRATION

Moths include several of the most notable and economically consequential insect migrants. A significant fraction of the world's most destructive crop pests migrate on storm fronts, sometimes in hordes so massive that they show up on weather-tracking radar. A widespread migratory scenario among moths is to move out of tropical and semitropical areas into temperate regions to exploit the abundant, nutrient-rich vegetation that becomes available each spring. In tropical regions, analogous mass movements take place between wet and dry forests that are synced up with rains and allow the moths to take advantage of the availability of new growth.

Because these movements occur at night, the migrations of moths are less familiar and less studied than those of butterflies. There is also increasing evidence that many, if not most, moths migrate at heights of 1,000 ft (300 m) or more—

BELOW | Urania Swallowtail Moths (*Urania fulgens*) puddling during mass migration to a forest with new growth of *Omphalea*, its larval host.

ABOVE | Grizzly Bear flipping rocks in search of aggregations of aestivating Army Cutworm (*Euxoa auxiliaris*), which make up an important part of the summer diet of Grizzlies in the Rockies of North America.

moving on aerial conveyor belts that can propel their movements at speeds often exceeding 60 mph (100 kmh). At these velocities, moths, and especially the microlepidopterans, are essentially aerial plankton that are simply taking advantage of atmospheric currents that will carry them great distances over a single night.

Sunset moths (family Uraniidae) make spectacular migrations. These swallowtail-sized, day-flying moths—dazzling in their beauty of intermixed metallic green and orange scales—are renowned for their mass movements from jungles of older foliage to sites in which their host plants are producing new leaves. While most of the migration occurs above the canopy, and might be entirely missed by the casual naturalist, both sexes become visible at ground level when they descend to gather water and salts at mud.

Two related noctuids warrant special mention: the Army Cutworm (*Euxoa auxiliaris*) and the Bogong Moth. The Army Cutworm is an abundant species that inhabits the high prairies, grasslands, and steppes flanking the Rocky Mountains of the United States and Canada. Its larvae mature on the lush grasses and forbs of spring (including alfalfa and row crops). The adults eclose weeks later by the millions, and begin their annual migration into the mountains, where they gather in huge numbers in talus slopes and aestivate over the dry summer months, when their grassland habitats have browned and become hostile. Millions if not billions remain in a state of reproductive diapause, sequestered under rocks in the high scree. In Yellowstone National Park, and no doubt elsewhere, many insectivorous mammals dine on the aggregations through the summer months. Most famous among these are grizzly bears. A single bear may eat 20,000–30,000 moths a day—as much as one-third of the calories required for an entire year may derive from the consumption of the moths. With the return of rains in the autumn, the survivors emerge from their aestivation sites and begin the reverse migration to the greening grasslands where they will lay their eggs and start the cycle anew.

Australia boasts an even more exceptional migrant, the Bogong Moth. Its biology mirrors that described above of the Army Cutworm, being a grass and forb feeder whose caterpillars mature in spring in low-elevation grasslands. Until recently its numbers were estimated to be in the billions, and it ranked among the world's more abundant animals. Dozens of mammals and indigenous nations fed on the aggregations of aestivating moths that gathered in caves across southeastern Australia (see also page 42). During the spring migration the moths were so numerous as to slow trains and shut down many public events. The fortunes of the Bogong took an abrupt turn in 2017 when a severe, three-year drought struck eastern Australia. Moth numbers suffered mightily, many caves have since gone unoccupied, and the Bogong is no longer seen in abundance over much of its range. So grave has been the decline that the moth was listed in 2021 as an endangered species by the International Union for Conservation of Nature (IUCN).

The fraction of moths that migrate is poorly known, as virtually all the flight activity happens under cover of darkness, and much is far above the ground where the phenomenon is intractable to

LEFT | The mass migrations of the Bogong Moth (*Agrotis infusa*) were legendary—dense enough to close down schools and church services and disrupt travel.

study. As a general rule of thumb, migration is the norm in dry, seasonal habitats, deserts, and other arid lands, where water is scarce and vegetation becomes unsuitable for caterpillar development over long periods, and especially in regions where the rainfall pattern is sporadic. It is also common from tropical and subtropical regions that do not experience freezes, into temperate and boreal areas, in spring and summer.

DIURNALITY IN MOTHS

While many of the most archaic moth lineages are brightly colored day-active animals, the vast majority of moths are nocturnal. The five largest families of Lepidoptera, in order of decreasing richness (Erebidae, Geometridae, Noctuidae, Tortricidae, and Crambidae), are essentially nocturnal. Yet, each of these families has spawned multiple lineages that are day-active, but virtually all are small groups and of little ecological consequence. I suspect if we knew the phylogeny of moths in detail, we would learn of more than 200 instances where a nocturnal moth group gave rise to a new diurnal species or lineage, but only one of these has met with great success—we call these butterflies. Three moth families with a preponderance of diurnal species include the Castniidae, Sesiidae, and Zygaenidae.

Moths that inhabit cold environments—high latitudes and alpine ecosystems—are often diurnal, presumably because daytime temperatures are more favorable and vertebrate predation pressures are lowered. In alpine communities and above 60° north or south latitude, a great many moths are diurnal perforce. Several traits are associated with this transition.

LEFT | Diurnal moths often have comparatively large compound eyes, and none more so than those of fairy moths, where the eyes sometimes fuse over the top of the head.

ABOVE | Aposematically colored moths typically are rendered in bright red, orange, and yellow, often paired with black or white markings, to warn of their unpalatability.

With the exception of butterflies, most diurnal moth lineages are smallish, so small that they are commonly ignored by birds and other visual predators. Nocturnal species tend to be green, brown, gray, or rendered in earth tones that can go unnoticed during the day when they are perched. By contrast, diurnal moths are often white, brightly colored, or otherwise more conspicuously rendered than their nocturnally active sister taxon. Diurnal moths that are palatable tend to be fast fliers, making them challenging quarry.

A large fraction of diurnal moths includes those that are chemically protected and unpalatable to birds. As such, they can operate in daylight with some level of impunity—these tend to be aposematic (brightly colored), slow-flying, and often hardy in constitution, that is, potentially capable of surviving a predator attack. Such species often anchor the abundant mimicry systems (see page 62) found across the order.

Because visual cues replace the primacy of odor communication in day-flying moths, the antennae are often smaller than those of closely related nocturnal species. In many moths the compound eyes of day-active species may be larger, and especially so in male diurnal adelids, with enormous compound eyes, that may join over the top of the head. But in still other lineages the eyes of diurnal moths are conspicuously smaller than those of their closely related nocturnal cousins.

DISTRIBUTION AND HABITAT

Moth diversity increases with proximity to the equator, although there are important lineages that

LEFT | *Euryglottis aper*, a large tropical hawk moth of western South America. Note its large, coiled tongue, which extends for many inches when it's actively nectaring (pollinating).

ABOVE | *Automeris* is a large New World genus with more than 70 species. The prominent eyespots are concealed at rest. Upon disturbance, the forewings are thrown forward to startle its attacker. Shown here is *Automeris amanda*.

become more speciose at higher latitudes (before dropping off at still higher latitudes). It is possible that as much as 80 percent of all moth diversity is endemic to tropical ecosystems, with diversity peaking in the neotropics. Nowhere is as rich as the foothills of the Andes, with the equatorial regions of Colombia, Ecuador, and Peru boasting the highest planetary diversity of moths.

Moths occupy virtually all terrestrial communities; if there are plants, there are likely to be moths. A few lineages are fully aquatic in freshwater—with the acentropine crambids being the most diverse and ecologically successful. Acentropine caterpillars feed on algae as well as aquatic plants. As would be expected, moth richness increases with plant diversity: scrublands have more diversity than grasslands and woodlands support more species than shrublands. Likewise, moth diversity increases with the architectural complexity of their hosts: trees support more species than shrubs, which support more diversity than forbs. In temperate regions, it is likely that the 10 most abundant tree genera support more than half of all the species of moths in a given community. Across the northern hemisphere, but especially in North America, oaks are the clear frontrunner: more than 1,000 species of moths are known to feed on oaks in America north of Mexico—a number that will increase as life histories of many western moths still await discovery.

NATURAL ENEMIES

Moths and their caterpillars are eaten by legions of other animals both small (for example, ants) and a sweep of larger animals: fish, amphibians, reptiles, birds, and mammals. Many birds are reliant on caterpillars, timing their migrations

ABOVE LEFT | *Eupackardia calleta* caterpillar. The larvae are disruptively colored as well as boldly marked. Upon disturbance, they will secrete a fluid rich in biogenic compounds from their scoli, which repels ants and other enemies. Note the very small clear droplets on segments A3, A7, and A8.

LEFT | Wasp moths have radiated across the planet's tropics. More than 3,000 species have been described. This Neotropical *Cosmosoma* is best regarded as a Müllerian mimic of wasps, as it is unpalatable as well, protected by the pyrrolizidine alkaloids consumed during its larval stage.

ABOVE | (*Left*) A swarm of aquatic moths. (*Right*) An early instar aggregation of Buff-tip moths (*Phalera bucephala*) on an oak leaf.

and breeding to annual peaks of caterpillar abundances in late spring, with both clutch size and fledging success often tied to caterpillar availability. One study found that it took 6,000–9,000 caterpillars to rear one clutch of Black-capped Chickadees. Lizards and snakes are also avid caterpillar hunters.

Caterpillar- and pupa-feeding mammals include bats, mice, voles, shrews, chipmunks, squirrels, raccoons, skunks, foxes, and bears. In tropical regions, monkeys are important predators of caterpillars. Many indigenous peoples also eat caterpillars (see pages 41 and 42).

Adult losses to vertebrates pale in comparison to those suffered by the eggs, caterpillars, and pupae. However, bats harvest enormous quantities of moths. The Mexican Free-tailed Bat colony of perhaps 1.5 million adults that roosts under the Congress Avenue Bridge in Austin, Texas, is estimated to consume over 20,000 lb (9,000 kg) of insects most nights when pups are nursing—much of that harvest is moths. Nighthawks and related caprimulgid birds are also moth specialists.

Invertebrate predators account for most caterpillar predation. Foremost among these may be ants—especially in tropical and semitropical regions, where ants may exceed the mass of the resident vertebrates. For example, more than 400 species of ants have been recorded from La Selva Biological Station in Costa Rica—most of these eat caterpillars. Then come spiders. Lynx and other foliage-gleaning spiders are important enemies of caterpillars. Large orb-weavers build efficient moth-trapping webs and even have adhesives that appear to be specialized for the capture of moths. A few spiders use analogs of a moth's mating pheromone to attract males that are in search of calling females. Bolas spiders spin a silk line, with a sticky terminal droplet laden with an attractant, which they wield about with a leg, to ensnare incoming males. A few species of large, orb-weaving spiders of genus *Argiope* produce volatiles that attract mate-seeking, day-flying saturniids into their webs.

Yellow jackets and paper wasps are especially fond of caterpillars and spend much of their

LEFT | A mongoose carts off a saturniid caterpillar. Diurnal mammals and birds that hunt caterpillars and other insects by day have shaped everything a caterpillar is about: its form, its color, how it feeds, when it feeds, and more.

BELOW LEFT | Spiders take a major toll on caterpillars, especially through early instars when the larvae are small. Later, they thin the ranks further, trapping moths in their webs.

foraging time hunting them. To these, we can add the legions of earwigs, assassin bugs, stink bugs, lacewing larvae, lady beetles, and other lineages of wasps. Some sphecid wasps provision their nests with caterpillars that they have paralyzed (but not killed, as these would soon rot). Entombed within the wasp's underground nest, the hapless caterpillar is then slowly eaten alive by the larva of the wasp. Caterpillars that feed on forest-floor plants, or descend to the ground to pupate, fall prey to still other enemies. Wolf and more than a dozen other families of spiders and ants likely get the lion's share, but centipedes and ground-dwelling beetles also pose a threat. Much of the text above addresses predators and parasitoids that attack middle and late instars. However, in many species, the greatest mortality rates occur in the egg and in the early instars, where the ranks are thinned by predatory mites, minute wasps, lacewing larvae, lady beetles, and other foliage-gleaning predators.

With the exception of mites, punkies (minute flies), and a few minute parasitic wasps, neither moths nor their caterpillars have any true parasites—that is, small animals that feed on them and then move off without killing the host. However, the early stages of moths are targeted by a seemingly endless number of flies or wasps that attack the eggs, caterpillars, and pupae. There is growing molecular evidence that there may be more than 20,000 species of parasitoid flies (nearly all Tachinidae) and more than 400,000 species of parasitoid wasps with most of these targeting the early stages of moths. At least three families of minute wasps, but especially trichogrammatids, attack insect eggs. Most parasitoid wasps and flies specialize on the larvae and pupae of Lepidoptera, with tachinid flies, and braconid, chalcidoid, and ichneumonid wasps being most prevalent.

The vast majority of insect parasitoids feed internally. Parasitized larvae suffer rather gruesome deaths. Typically, the larval stage of the fly or wasp feeds initially on nonlethal tissues, and then, in a final pulse of growth, consumes much of the caterpillar. In most, the host caterpillar is killed at the time the parasitoid matures and exits the vanquished cadaver. In some, the caterpillar, while mortally compromised, is fated to stand guard over the cocoon(s) of the wasps that attacked it, until the new generation of wasps hatch and fly off. A few groups of ichneumonid wasps are specialized on moth pupae. These can be observed hunting for cocoons and pupal cells on forest floors. Caterpillars and pupae in moist environments—marshlands, mesic forest, and along riparian corridors—also fall victim to both nematodes and horsehair worms.

The same types of pathogens that infect most animals also attack moths and caterpillars: viruses, bacteria, protozoans, and fungi. The fungus *Cordyceps*, an enormously successful genus with perhaps 600 species worldwide, attacks many Lepidoptera, especially those that live in soil as larvae. *Beauvaria* fungi, too, are a common enemy of caterpillars, especially those that live in the soil or other moist environments.

Many of these disease agents have been used in biological control programs to control pest species, in part because pathogens tend to be very specific

ABOVE LEFT | More than 8,500 tachinid flies have been described, with thousands more awaiting recognition—most of these parasitize caterpillars. *Compsilura cocinnata* (shown here) was introduced into North America as a biological control agent to attack the Spongy Moth (*Lymantria dispar*).

LEFT | Tiger moth caterpillar attacked by a microgastrine braconid wasp. More than 30 of the wasp cocoons are visible here—and all have hatched (note the open or missing opercula).

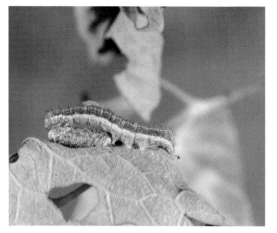

ABOVE LEFT | Zombie caterpillar. Having fed inside the noctuid, a microgastrine wasp larva exited and spun its cocoon under the caterpillar, which, still alive, serves to protect the wasp from predators.

ABOVE RIGHT | Pupa of a hepialid attacked by a *Cordyceps*. The fungus releases spores to infect the next generation of caterpillars.

and are only capable of infecting a small group of species. Extended periods of cool, wet weather, especially through spring months, favorable to pathogens, are associated with taxonomically widespread population downturns in many Lepidoptera. The fungus *Entomophaga maimaiga* has been spectacularly successful in bringing down outbreaks of the Spongy Moth (*Lymantria dispar*). The bacterium *Bacillus thuringiensis* is used worldwide in gardens, croplands, and forests to control pestiferous caterpillars. In general, viruses have highly specific host ranges, but are expensive to produce, and as such are only rarely employed as biological control agents. However, entomopathogenic viruses have great potential to be used in genetic engineering—for example, when their toxin-producing genes are inserted into a plant genome.

APOSEMATISM AND MIMICRY

While the vast majority of moths are cryptic in coloration, rendered in camouflaging greens, grays, and earth tones that blend in with foliage, bark, or soil by day, moths that are chemically protected and unpalatable are commonly aposematic. That is, they advertise their presence with bold wing and body colors: yellows, oranges, and reds that are accentuated with additional white and black markings. Bright white colorations, which are conspicuous both day and night, are another way distasteful moths commonly advertise their chemical protection.

The chemical ecology of such moths is worthy of its own book. The toxins and defensive chemicals that ward off would-be predators—iridoid glycosides, cyanogenic glucosides, cardiac glycosides, and a sweep of alkaloids—consumed by the caterpillar, sequestered and concentrated, can be passed through the pupal stage and on to the adult. Few moths manufacture their own defensive compounds: these are usually simple acids, aldehydes, and ketones. Smoky moths (Zygaenidae) and related families are famous for their ability to manufacture cyanide or compounds that yield cyanide when either the caterpillar or adult is under attack.

Typically, the defensive chemicals occur in high concentration throughout the body but may also be incorporated into wing and body tissues. A special case is that of *Hylesia metabus* of northern South America. The hairlike deciduous scales of the female's abdomen are highly irritating to human eyes and skin, and can be so problematic during periods when the adults are common that villagers will turn off their lights in order to protect themselves and their homes. A few lineages exude toxin-laden hemolymph ("blood") when under attack that can, in some cases, terminate the attack and allow the moth to escape.

Mimicry, arguably one of the most compelling cases of evolution and the uncanny powers of natural selection, was unknown to Darwin at the time he wrote *On the Origin of Species*. Instead, English naturalist Henry Walter Bates discovered mimicry while studying butterflies in South America. Given that butterflies are little more than day-flying moths, it should come as no surprise that many moths are mimetic, especially among the day-flying species, where color and patterning are crucial for survival. Moths may be

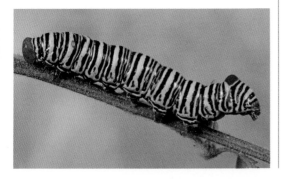

ABOVE LEFT | Unpalatable caterpillars (here, *Didugua argentilinea*) warn of their unpalatability with bright colors, patterns, and conspicuous behaviors; for example, they are much more likely to eat during daylight hours.

LEFT | Tiger moths may represent the largest radiation of mostly aposematic animals on the planet, with more than 11,000 described species.

either Batesian (where palatable species mimic a toxic model) or Müllerian (where unpalatable species come to resemble one another). Among these are hundreds of bee and wasp mimics, which have clear wings, devoid of scales, that resemble bumblebees or wasps in flight. The likeness of some clearwing moths (Sesiidae) and wasp moths (ctenuchine Erebidae) to their models is so close that all but the best-informed insect aficionados are likely to be fooled. Add to these a scattered smorgasbord of small, day-flying cossids, saturniids, sphingids, lasiocampids, a smattering of zygaenoids, and still others.

LEFT | The deciduous body hairs and scales of *Hylesia metabus*, a diminutive saturniid endemic to northern South America, cause particularly aggravating cases of region-wide dermatitis when they are on the wing.

BELOW | This ostentatious pericopine tiger moth (*Composia credula*) is protected by high titers of pyrrolizidine alkaloids (PAs) that are sequestered by its caterpillar.

OBSERVATION

Moths are among the easiest insects to observe as the vast majority come to lights at night, especially light sources rich in UV wavelengths. Partly for this reason, moth watching and especially moth photography are rapidly gaining in popularity among naturalists and community scientists around the world. Dozens of social media groups now anchor to moth watching and photography.

WHERE TO FIND MOTHS

The first rule of thumb is to search for moths at sites with high plant diversity. While woodlands and forests generally have the greatest species diversity, ecotones where early successional habitats intermingle about forested communities will be the most productive. Over the course of time, seek out different plant community types. Where water is limited, explore canyons and sites with water at or near the soil surface. Diversity drops off quickly with human activity: target sites away from yards, artificial lighting, and areas of abundant exotic plant growth, when circumstances allow.

Expect substantial species turnover during the year, with richness peaking for adults in late spring in temperate areas, and at the start of the wet season in ecosystems with a pronounced dry season. In extremely wet forests, species richness may peak in the dry season. For caterpillar hunting, shoot for three weeks after the peak for

BELOW | The White-lined Sphinx (*Hyles lineata*) is one of the most important pollinators across the deserts and drylands of North America.

LEFT | Meadows rich in floral resources are a great place to search for moths by day and night. Dusk is a special time that can be particularly rewarding.

LEFT | The Madagascan Moon Moth (*Argema mittrei*), regarded by many to be among nature's most beautiful and extraordinary animals. Tail lengths vary considerably—this individual is especially well endowed.

adults. Different species fly at different times of the year—many fly only in spring, others in summer, some lineages eclose in the fall with a subset of these persisting through the winter as adults—these may be seen on warm winter nights feeding on sap flows. This latter set tends to go into reproductive diapause until late winter and early spring, when mating occurs.

Different species have their peak flight at different times of the night or day. Dusk is a wonderful time to hunt for moths as many species are on the wing shortly after sundown when there is ample light to see without employing a headlamp. This is a prime time to look for many ghost moths (Hepialidae), plume moths (Pterophoridae), and other microlepidopterans.

Many giant silk moths (Saturniidae) fly principally after midnight. The twilight period before dawn is supposed to be especially good for grass-miner moths (Elachistidae), but is relatively quiet, with most moths making beelines for safe resting sites to pass the day.

LIGHTING FOR MOTHS

While moths will come to any bright green, blue, or UV-rich light, UV wavelengths are the most attractive, and a must for the serious moth photographer or collector. You can enhance a light's effectiveness by placing the bulb proximate to a white sheet that reflects much of the light out into a habitat that is the target of a night's sampling efforts. Within a night or two, stationary lights become feeding stations for bats by night and birds by morning. Shut your lights down as much as an hour before dawn and shake the sheets to give the moths time to settle elsewhere, especially if you are running a light at the same location night after night.

Many light trap variations are available for use with or without killing agents. Live trapping is highly encouraged where the moth fauna is known, imperiled species are present, where images of live moths are desired, and when seeking to capture a gravid female for breeding purposes. One live-trapping method that is both inexpensive and effective is to place the light among or directly over an assortment of 10 to 20 egg carton tops or bottoms that provide numerous

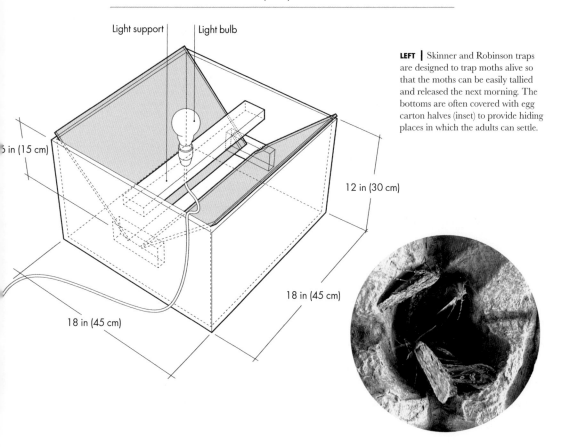

LEFT | Skinner and Robinson traps are designed to trap moths alive so that the moths can be easily tallied and released the next morning. The bottoms are often covered with egg carton halves (inset) to provide hiding places in which the adults can settle.

RIGHT | Sugary baits that have begun to ferment are highly attractive to many moths and are a reliable means to see some moths that are only weakly attracted to light; migratory erebids sometimes gather in huge numbers; nearly all noctuine winter moths are bait-feeders; virtually all *Catocala* (shown here) are drawn to fermenting baits.

dark places in which moths can take shelter. Place the set of egg cartons in a wide-mesh laundry bag or the equivalent (see page 70). Live traps should be serviced in the morning, before the traps are exposed to direct sunlight, with all but the needed species released.

Scientists use kill traps when they have minimal opportunities to visit a site over the course of a year—for example, when sampling in remote locations, when moving between sites on consecutive days, when a site is due for development or about to be lost to agriculture, for scientific purposes, and other reasons. Their use should be judicious and efforts made to ensure maximal use of the collected insects. The Lepidopterists' Society has a "Statement on Collecting Lepidoptera" that should be followed by professionals, students, wildlife biologists, environmental consultants, and others.

BAITING

Males and females of many moths take nourishment from the sugary solutions provided by flowers, broken tree limbs, oozing plant wounds, fruits (including those that are overripe or even rotting), as well as accumulations of honeydew excreted by aphids and other homopterans. Tree wounds are especially attractive to many moths—any moist bark patch that has an abundance of flies, wasps, and especially butterflies during the day is sure to be a flurry of moth activity by night. These same moths can be drawn to sugary baits that are fermenting, with those that smell strongly of alcohol performing best.

Bait concoctions vary from simple to complex, and even bizarre. Simply "paint" it onto a tree trunk, at about chest level, or place it in a bowl elevated above the ground. In treeless landscapes,

ABOVE | Moths make wonderful subjects for backyard and local-area surveys and offer a gateway to the study (and protection) of nature.

the bait can be poured over sponges placed on inverted lids or some other solid substrate.

Like mammals, moths sometimes become intoxicated—species that are wary and unapproachable under normal conditions can be so intoxicated that you can pick them up by hand and photograph them. Conversely, underwing moths (Erebidae: *Catocala*), the principal quarry of many moth baiters, typically remain wary and will flee as soon as a strong light is shown on their position. Approach baiting stations quietly and avoid shining bright lights directly on the feeding moths. The red wavelengths of some headlamps are less alarming, but you still want to avoid shining the light directly on the moth. Many moths have ears—a stealthy approach without conversation is best.

Sugary baits work well in the fall and winter for myriad erebids, noctuids, and some lineages of tortricids and oecophorids. The method works best where sugary resources such as overripe and rotting fruits are commonly available to the local moth fauna; conversely, baits are seldom effective in desert and dryland habitats that lack naturally occurring sugary analogs to moth baits. Expect the yields to differ greatly and inexplicably among nights, depending on the bait, humidity, recent rains, cloud cover, and ambient weather conditions. Baiting can be especially productive over droughty periods, when natural alternatives have been scarce.

Finding and photographing diurnal moths is more challenging, and in many ways more rewarding, as you often have to know more about

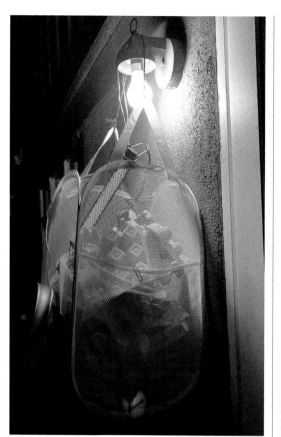

LEFT | One of the easiest means to live-trap moths is to place a UV light source and egg cartons into an open-mesh laundry bag. Moths attracted to the light will settle into cartons where they are protected from birds and easily examined the next morning.

the moth's biology, preferred habitat, host plants, nectaring preferences, and diel behavior to be successful.

Hunting and raising caterpillars of moths can be an engaging hobby, providing wonderful photographic opportunities, and offering much potential to yield new observations and discoveries. Such is especially true in the tropics and many regions of the southern hemisphere. Caterpillars are relatively easy to raise. Any such efforts are most valuable when you can confidently identify wild host plants; secure quality images of the early stages; photograph any issuing adults; and share your findings. Such life history data are lacking for most species of moths, and are critical to conservation and restoration efforts.

MOTH GARDENS

Butterfly and pollinator gardens are becoming exceedingly popular in home, community, nature center, and school gardens, and additionally are serving to drive interest in insect conservation matters more broadly. The idea of planting a garden for moths—for visitors that benefit after most people have gone in for the night—has only recently gained traction. While many good butterfly flowers, such as lantanas, buddleias, and milkweeds, are also good for moths, other moth flowers differ from those recommended for pollinator gardens. Moth flowers are often white and quite fragrant—think jasmine and honeysuckle. Many have a deep corolla tube that hides the nectar at the bottom so that only long-tongued moths, such as sphingids and noctuids, can easily access the nectar rewards. Various phlox, campions, and verbenas are examples. Four-o'clocks and evening primroses may not open their flowers until late afternoon or twilight (respectively), with moths clearly intended as their preferred pollinators. In addition to targeting the visitation of the adults, consider planting the larval hosts in insect gardens.

CONSERVATION

The International Union for Conservation of Nature (IUCN) maintains a list of imperiled species of Lepidoptera, but from its inception the list has been focused on butterflies. There is much interest in ramping up global efforts to evaluate the conservation status of moths, especially those endemic to imperiled habitats and those threatened by climate change. Readers are encouraged to assist the IUCN's effort by sharing their occurrence data, actively participating in status surveys and listing efforts, and by otherwise contributing to efforts to conserve moth biodiversity.

EXTINCT, ENDANGERED, AND THREATENED SPECIES

Fewer than 150 species (out of more than 160,000 described species of moths) have been evaluated by the IUCN. Thirty-five moths are listed as critically endangered, 21 as endangered, and 11 as near threatened. Forty-five percent of the imperiled species are from Hawaii (Oceania)—one of the most threatened ecosystems on the planet—where the fauna has been ravaged by exotic plant introductions, introduced ants and parasitoid wasps, and, on the bigger islands, urbanization. The IUCN lists 17 extinct species of moths—11 of these are from Hawaii.

The Palearctic realm is home to the best-known continental moth fauna and the western part of the region is currently the focus of a separate European Red List Project. The IUCN database for the Palearctic includes only eight critically endangered, eight endangered, and four near-endangered moths. None of the approximately 25,000 species of moths recorded from this realm are known to have gone extinct.

The incompleteness of our global lists cannot be overemphasized. There are more than 300 additional moths from Hawaii that have not been seen in more than a half-century. Island faunas around the world, especially those that are remote from the mainland and rich in endemic taxa, are threatened by invasive species, habitat destruction, and agricultural practices. Invasive ants are among the most pernicious of all exotic

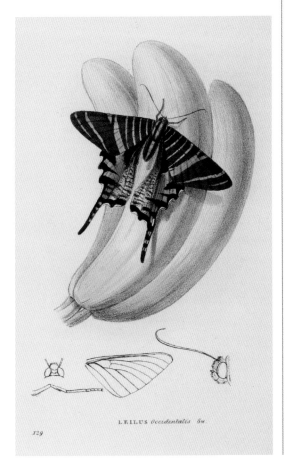

LEFT | *Urania sloanus*, endemic to Jamaica and one of the world's most beautiful moths, may be the first moth known to have gone extinct. It was last seen in the late 1800s, presumably falling victim to deforestation and agriculture.

DEATH BY A THOUSAND CUTS—GLOBAL THREATS TO INSECTS

Global Warming
Arctic sea ice is declining precipitously; Arctic–alpine and other cold-adapted communities are contracting; sea-level rise threatens coastal ecosystems.

Fire
Wildfires are becoming more frequent and intense, and account for nearly twice as much habitat loss as 20 years ago.

Droughts
Periods with diminished precipitation are becoming longer, more frequent, and warmer, with grave consequences for all life.

Storm Intensity
Climate changes bring: stronger, more frequent storms and hurricanes; more fire-igniting lightning; and damaging flooding.

Interaction Disruption
Climate change is affecting ranges of many species globally, with expansions and contractions disrupting biological interactions established over the Holocene.

Deforestation
25 million acres (10 million ha) of forests are cut or burned annually, mostly for agriculture, releasing 4.8 billion tons (4.35 billion tonnes) of carbon dioxide into the planet's atmosphere.

Agricultural Intensification
Industrialized agriculture, with its large increases in scale, monoculturalization, nutrient input, and pesticide use, is becoming ever increasingly nature unfriendly.

Nitrification
Fertilizers and products of fossil fuel combustion nitrify the planet, especially challenging the biotas adapted to low-nutrient conditions.

Pollution
Chemical, light, and sound pollution of water, air, and soil are impacting plant and animal life worldwide.

Urbanization
Our global population of 8.2 billion comes at great cost to biodiversity and wild lands. Already, over 500 vertebrates have been driven to extinction.

Introduced Species
Global trade is accelerating the movement of pernicious plants, animals, and pathogens to new regions, often with devastating consequences

Insecticides
Modern agriculture, with its increasing reliance on chemical insecticides, has led to chronic contamination of wild lands, and impacts nontarget insects.

THIS PAGE | The figure depicts a dozen of the principal stressors acting on insects globally. Others could be added, for example, the increasing variability in global climates, which disrupts insect life cycles. The impact of car strikes by day, and by night (especially for moths), is not insignificant. Adapted from Wagner *et al.* (2021).

ABOVE | On Barro Colorado Island in the Panama Canal, far removed from agriculture and deforestation, many insect populations are stable.

introductions as many island biotas (including those of Hawaii) evolved in the complete absence of ants; but, once established, their numbers can explode, with virtually every native insect becoming vulnerable to their maraudings. On some islands, biological control agents, insect parasitoids and predators, have been purposefully released to control pests on agricultural crops, with calamitous consequences for the nontarget, native moths.

INSECT DECLINE AND THREATS

Reports of a loss of insect diversity have been made from the poles to the equator, with the steepest rates documented from Europe and North America. Indeed, much of the evidence for "insect declines" comes from population trend data for butterflies and moths, which are among the most familiar and frequently sampled insects. Two ecosystems in which steep declines have occurred include arid regions of the American Southwest and the tropical dry forests of northwest Costa Rica. Moths are not declining everywhere, however. Many species are increasing in range in temperate regions where their distribution was previously constrained by cold winter temperatures. Across western and northern Europe, collectors and moth watchers are recording hundreds of species of moths that have extended their ranges northward. Moth populations appear to be stable or increasing on Barro Colorado Island situated in the Panama Canal, where the moth fauna is buffered from deforestation, agriculture, and other land use changes, as well as pesticides, pollution, appreciable invasions of exotic plants, and, so far, detrimental changes in climate. By contrast, in other tropical areas, especially those proximate to cities, agriculture, or where forests are aridifying, moth numbers appear to be in steep decline.

HABITAT DESTRUCTION

The single greatest stressor of invertebrates globally is thought to be land use change, a collective euphemism for deforestation, conversion to agriculture, commercial and industrial development, and urbanization and exurbanization. Of these, the first two are overwhelmingly among the most important drivers of biodiversity losses. Each year 24.5 million acres (10 million hectares) of forests are cut, with nearly half of this area converted into agriculture, residences, and other human uses. When forests, woodlands, and grasslands are reduced to croplands, the consequences are especially costly for nature. Where natural areas are replaced by a sea of just one or two crops, much of the former ecological heterogeneity is reduced to an edaphically and hydrologically simplified landscape, onto which an unending litany of fertilizers, herbicides, and insecticides will be applied. Worse, agrochemicals invariably leak into and fundamentally challenge adjacent communities. Many of the highest rates of well-documented insect declines come from natural areas adjacent to croplands.

LIGHT POLLUTION

Light pollution is being increasingly recognized as a threat to moths and other species attracted to lights. Artificial lights serve as feeding stations for insectivorous bats by night and birds the next morning—add to these the maraudings of predaceous beetles and ground-feeding rodents that use permanent light sources as hunting grounds. Moths that circle, round and around, under a light source waste tremendous energy and food reserves. Such is especially harmful for

BELOW | A water-starved lowland, tropical thorn scrub community in northwestern Costa Rica, suffering from an extended drought and higher than normal temperatures.

taxa that lack a haustellum and functional digestive tract, thus lacking any chance to refuel. Moths are in decline where artificial lights are used in abundance, for example, where porch, yard, and street lights are left running night after night over long periods.

CLIMATE CHANGE

Many biologists think that climate change is overtaking habitat loss as the greatest driver of population and species losses of moths and other biodiversity. This is especially true in polar regions and alpine ecosystems where average temperatures are changing rapidly; in coastal communities that are subject to storm surges and flooding; and in deserts and arid lands where droughts and elevated temperatures are becoming more frequent and of longer duration. How climate is affecting the world's tropics, where as much as 80 percent of the world's insect diversity is found, is of particular concern and remains a huge data gap. Recent precipitous declines of moth numbers in northwestern Costa Rica appear to be linked to droughts and the prolongation of the dry season, more frequent and severe droughts, climate variability, and especially more erratic starts and stops to the annual dry and wet seasons. The American Southwest is suffering from the worst drought in the last 1,200 years—moth numbers and diversity are down across the region.

THINGS YOU CAN DO

There are many immediate and long-term activities that will help moth populations. Turn outdoor lights off when not needed and replace incandescent and halogen bulbs with LED bulbs; use bulbs that emphasize yellow wavelengths that are less visible to moths; shield outdoor lighting so only essential areas are illuminated; and, most importantly, install motion-detecting lighting systems whenever possible. Stop or significantly reduce using pesticides and fertilizers for lawns and other cosmetic purposes. Plant a pollinator garden that emphasizes moths' preferred nectaring sources and larval host plants.

In recent years, photographing moths, caterpillars, and other insects has become an enormously popular pastime across all six of Earth's inhabited continents, and has grown to such a popular hobby that it is laying the foundation for social networks, research collaborations, technique development, trip planning, and more. Phone apps and websites anchored to moth photography are gaining exponentially in popularity, and surely represent an important development in changing public attitudes about moths. At the same time, many facilitate nearly immediate identification, store and serve the collective data, and emphasize the building of online communities anchored to moth biodiversity. Phone apps harnessing AI technology to suggest species-level identifications are improving by the month and do quite well for many countries.

Efforts to educate and involve others about the importance, beauty, fascinating life histories, and opportunities for discovery of moths will generate interest and likely willingness to take actions to protect moths and other invertebrates. Starting with young children, activities in and out of the classroom will spark positive memories and long-term rewards. In particular, those activities associated with museums, nature centers, and society events are wonderful. One of the most impactful means of generating interest in moths and their caterpillars is getting others involved in raising caterpillars—observing metamorphic changes over the course of two to three weeks can be life changing.

LEFT | Butterfly houses and insect zoos have become popular destinations for school groups and public visitation. They play important roles in changing perceptions about moths and other insects.

REWILDING AND MANAGEMENT

There is much to recommend in terms of efforts to preserve and restore habitat for insects—in our yards, town common lands, parks, and especially along corridors that facilitate movements between reserves. It can be enormously satisfying to rewild sections of a lawn with wildlife habitat and replace exotic plants with natives. There are dozens of examples where small, well-managed habitats have been highly successful. Two aspects of moths and other insects make them good targets for such efforts. One is that some species can persist or be rehabilitated on very small pieces of land. The second is their tremendous capacity to increase in number. Many insects routinely have boom-and-bust life cycles, in which population numbers of adjacent generations can vary by an order of magnitude or more. Monarch butterfly numbers in the western North American overwintering aggregations in California went from an all-time low of 2,000 butterflies in 2019 to more than 200,000 butterflies just two years later.

While there are advantages to working on insect conservation, there are also challenges. The first of these, which might explain why there are so many opportunities in invertebrate conservation, is that there are at least 20–30 times as many insects as there are vertebrates; but, paradoxically, less than one in twenty dollars spent on conservation goes to invertebrates. Another challenge of insect conservation is that many species require managed, early successional habitats that take solid science and active management to maintain in perpetuity. Insects are poikilotherms that require sun to warm their bodies, especially across large swaths of the temperate zone. A great many thrive in early succession habitats—fields and meadows, sand bars and flood-prone bottomlands, grasslands, meadows, sandplains, shrublands, fire-adapted communities, and other open, sunny habitats. Virtually all of these are dependent on ecological disturbances. Such habitats require active management to maintain, and thus require long-term commitments from the conservation community or dedicated individuals.

LEFT | Insect walks and related activities are enjoying heightened interest worldwide, driven in part by cell-phone technology, greater macrophotography options, and new identification resources that allow even the uninitiated to acquire quality insect imagery and immediate identifications.

MOTH COLLECTING

Collecting is both encouraged and discouraged. Creating and maintaining a well-curated scientific collection, where information and specimens are shared, has much to recommend it as a pastime, means of training, and as a career path for those interested in pursuing a professional position in systematic entomology. But collections must be made in an ethical and responsible manner.

Properly conducted scientific collections with well-preserved voucher specimens are of great scientific and conservation value, especially for areas and taxa that are in need of biosystematic study or habitats that will be lost to development or agriculture. Each voucher is a trove of DNA, contains tissues for toxicological study, includes the individual's microbiome and epifauna, and may retain pollen from recent flower visitations. Its internal structures—features of the male and female genitalia—are often needed for authoritative identifications. Well-curated specimens are the essential infrastructure needed for taxonomic studies, are ultimate vouchers for establishing identities in many scientific studies, and have especially high value in regions and countries where the fauna is still poorly studied and existing taxonomy is nascent. In the future, insect collections will have tremendous legacy value for many types of ecological and evolutionary studies, genomic research, morphological examination, phylogenetic inference, and certainly more.

While collecting might be an end for some, more are served if it is a beginning—that is, an opportunity to record behavioral and core ecological data, work out details of an unknown life history, gather population census data, and so on. While specimens from known areas have scientific value, those from new populations, habitats, and regions yield new knowledge and have an elevated scientific importance. Specimen vouchers deposited in a public institution coupled to scientific studies are a requirement for many scientific journals. Both larval and adult collections connected to life histories are particularly valuable—it is hard to know how to protect or manage habitat for a species until there is a solid understanding of the taxon's early stages, larval host plants, habitat preferences, and other life history data. Making collections of wild caterpillars and raising them to the adult stage can be enormously enjoyable and have great potential to yield new knowledge. For example, rearing wild collected caterpillars has the potential to reveal much about a species' parasitoids and associated hyperparasitoids.

Collecting for collecting's sake, without an understanding of scientific needs and potential consequences, is not condoned. Takes must be measured and never exceed 10 percent of a local cohort, and should be male-biased. Live trapping should be employed as much as possible. If a large series is desired, all are encouraged to embrace rearing efforts—that is, where adults are obtained from the rearing of eggs obtained from a gravid female. Commercial collecting, taking long series for trade, and profligate kill trapping where most of the catch is discarded, is not in the best interest of nature, and is now regulated in many countries.

How and why a collection is gathered, curated, maintained, and ultimately gifted to a public institution is straightforward: if nature is a beneficiary, it is a good collection and a greater good is served. If Mother Nature would not approve of your efforts—rethink your actions.

BELOW | *Manduca blackburni* is one of the many endangered species of moths endemic to Hawaii. There are more than 300 species of moth that have not been seen here in the last 50 years.

INTRODUCTION TO TAXON PROFILES

This book provides an overview of the currently recognized superfamilies of Lepidoptera. Due to a dearth of information on their biologies and their sparse species diversity, three superfamilies, Andesianoidea, Lophocoronoidea, and Simaethistoidea, are omitted. There is only space for 72 of the 128 recognized families of moths in this work, with preference given to taxa rich in species, of special evolutionary stature, economic importance, particular beauty, and those lineages of elevated conservation importance. Included among the latter are the nine early diverging lineages that are essentially relicts. These serve as windows to the deep past and genetically represent exceptionally novel entities. The number of pages devoted to a family is likewise a function of these same criteria, as well as wingspan, as the attention of most people is greatly influenced by an organism's size. In the taxon profiles that follow, attributes of forewing and hindwing are often indicated by FW and HW, respectively.

The increasing availability of molecular (DNA) data has spawned a renaissance in the taxonomy and classification of Lepidoptera. DNA sequence data are now routinely used to identify moths, circumscribe species, and infer the phylogenetic relatedness of taxa across the tree of life. DNA data will allow lepidopterists to infer when taxa originated and how they are interrelated—information that is essential to building stable classifications that reflect the evolution of life. Expect much change in the future, especially across the higher taxa that appear in this book: many subfamilies will be elevated to full family status (for example, the stenomines are regarded to be a full family by some); other taxa will prove to be subordinate to previously recognized groups (for example, the Dalceridae appear to have originated within the Limacodidae, and will need to be subsumed as a subfamily within the slug caterpillars).

The range maps are provisional, and based largely on records on iNaturalist, the Global Biodiversity Inventory Facility (GBIF), and the books and websites listed on page 234. The rapidly increasing global interest in moth watching and photography will soon correct this shortcoming, as moths, to a greater extent than most animals, will come to you, if you use a light, concoct baits, or employ synthetic pheromone lures (see page 36). And because of this, moths provide special opportunities for the young, aged, infirm, disabled, and those too burdened with responsibilities to travel afar. As such they serve as ready vehicles to explore nature's wondrous, endless invertebrate diversity and beauty.

It has only been possible to share a few images for any given taxon, and thus we are unable to do justice to the heterogeneity expressed across most moth families featured here. For virtually all the taxa, the reader will want to visit the additional resources mentioned on page 234 and look for books and websites treating local and regional faunas where you live or might be traveling. Given that we are looking down the front end of a global biodiversity crisis, many of the observations (think also discoveries) that you can make, even in your backyard or local park, will be a gift to posterity.

As made clear above, our knowledge of moths—their taxonomy, phylogenetic underpinnings, ranges, seasonality, early stages, host associations, behaviors, and more—is still in a nascent state, with much left for the weekend or backyard naturalist to contribute. Readers are especially encouraged to make efforts to learn more about our planet's poorly studied moths.

LEPIDOPTERA: MOTHS

While more than 160,000 species of Lepidoptera have been described, likely fewer than half have been given a name, and the larval stages are known for only about 20 percent. More remains unknown than known.

Lepidoptera and Trichoptera are sister taxa, whose common ancestor dates back at least to the Triassic. The oldest lepidopteran (moth) fossil dates to the middle Jurassic. Extant Lepidoptera share many characters that uniquely attest to their common ancestry. Most notable are their hollow, flattened scales, which also account for the boundless beauty and variety of moths. An especially perplexing attribute is their high percentage of anucleate (apyrene) sperm. In some moths, more than 90 percent of sperm lacks a nucleus. Why this is so and whether this has an important adaptive function has yet to be resolved. While unique to Lepidoptera, the haustellum is wholly absent in three early diverging superfamilies.

The profiles that follow are divided into three groups: non-ditrysian microlepidoptera, ditrysian microlepidoptera, and macrolepidoptera. The first heading for each profile gives the superfamily of the moths on the page. This is followed by the family name. For some larger groups, profiles are divided further into subfamilies and even tribes. (See page 14 for an explanation of the word endings associated with each group.) The larger gray heading is the commonly accepted English name for each grouping described.

OPPOSITE | Day-flying microlepidopterans include some of Earth's most beautiful insects, too often overlooked due to their diminutive stature. Shown here is *Hemerophila diva* (Choreutidae).

NON-DITRYSIAN MICROLEPIDOPTERA
ARCHAIC LINEAGES

This first section of taxonomic treatments anchors to the 13 superfamilies of non-ditrysian moths, which are informally referred to in this work as "the archaic" lineages, in that they diverged from other Lepidoptera early in the evolution of the order and retain many ancient anatomical and behavioral traits.

Adults of several non-ditrysian moths are small, brightly colored, diurnal moths, commonly with metallic gold or purple scaling, or both. The two most diverse families of non-ditrysians are night-active: Hepialidae (>600 species) and Nepticulidae (>850 species). But even when all 24 of the archaic families are pooled, they account for less than 2 percent of described moth species. While most are small moths with wingspans under half an inch (1 cm), the hepialids and andesianids often have wingspans in excess of 1 in (2.5 cm), with the largest having wingspans exceeding 10 in (25 cm). All but the Hepialoidea possess a single reproductive pore, for both copulation and oviposition. As is commonly the case among moths, those rendered in browns, grays, and other earth-tones tend to be primarily nocturnal.

The larvae of many non-ditrysians feed on ancient plant taxa—bryophytes (liverworts), ferns, and gymnosperms, as well as early diverging angiosperms. Hence, many members of these archaic lineages appear to be living in the same types of habitats and behaving in the same fashion as their ancestors!

The phylogenetic tree (opposite) shows evolutionary relationships for the early diverging superfamilies of

PHYLOGENETIC RELATIONSHIPS AMONG ARCHAIC LINEAGES OF LEPIDOPTERA

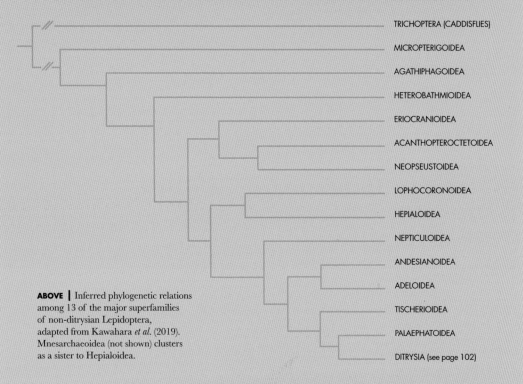

ABOVE | Inferred phylogenetic relations among 13 of the major superfamilies of non-ditrysian Lepidoptera, adapted from Kawahara *et al.* (2019). Mnesarchaeoidea (not shown) clusters as a sister to Hepialoidea.

Lepidoptera—most of these lineages are endemic to or much more speciose in Gondwanan regions of the southern hemisphere. As relict lineages, they offer a view back in time and rich opportunities for the study of Earth's deep history.

Adults of micropterigids feed on fern spores and the larvae of most are thought to feed on liverworts and even decaying wood and other forest-floor detritus.

Nine families of non-ditrysians are not featured, due to their rarity, poorly known biology, or because they are only infrequently encountered. While the oldest fossil moths date to the middle Triassic, 200 million years ago (mya), molecular data suggest that Lepidoptera diverged from caddisflies (Trichoptera), their sister group, sometime in the Permian.

The life histories of Lophocoronidae and four families of Hepialoidea are essentially unknown.

MICROPTERIGOIDEA: MICROPTERIGIDAE
MANDIBULATE ARCHAIC MOTHS

These colorful archaic moths are a must-see for many moth hunters. The more than 160 described species are placed into 21 genera, although many southern hemisphere taxa remain to be discovered and described. They are beautiful insects, with many rendered in metallic golds and purples, as is commonly the case among diurnal archaic lineages. A few are patterned in such a way as to be credible jumping-spider mimics.

Diurnal adults can be found perched on low vegetation in mesic forests, especially in the vicinity of sporulating ferns and where liverworts are abundant. *Micropterix*, which represents a disparate lineage from the main cluster of micropterigid genera, feeds on the pollen of wildflowers as well as on an array of woody plants. A few are nocturnal.

Larval biologies are poorly known. Most genera are thought to feed on liverworts. Green leaves of various flowering plants, detritus, and fungi are eaten by *Micropterix*. The pupa has enlarged, functional mandibles that are used to open the silken cocoon prior to adult emergence.

LEFT AND INSET | *Micropterix* accounts for almost half the described species of the archaic family. Adults commonly have shiny gold bands or spotting across the forewings. The caterpillars of micropterigids (inset) are so anatomically unique that at one time micropterigids were classified in their own order (Zeugloptera).

DISTRIBUTION
Cosmopolitan; more diverse in Old World

IMPORTANT GENERA
Micropterix, *Paramartyria*, and *Sabatinca*

HABITAT
Mesic forests and woodlands, especially of temperate zones

HOST ASSOCIATIONS
Liverworts, mosses, and some polyphagous on low vegetation and possibly leaf litter and other detritus

CHARACTERISTICS
• Wingspans from 0.3–0.5 in (8–12 mm), rarely to 0.6 in (15 mm)
• Antennae held upward and to side with some starfish-like sensilla
• Functional jaws; haustellum (proboscis) absent
• Mesotibial spurs absent
• Larvae subhexagonal or oval in cross section with relatively long antennae

AGATHIPHAGOIDEA: AGATHIPHAGIDAE
KAURI MOTHS

The two surviving species of these caddisfly-like moths are placed in a single genus: *Agathiphaga*. The adults are mandibulate, but it is not known how the mandibles are used; they may be used solely to free themselves from the hardened pupal chambers but also may be used for feeding on pollen, spores, or still other substrates, as in kindred families. The large crop of the adult foregut is suggestive that adults feed.

The brown and rather plain-looking adults are nocturnal and rarely encountered. Their grub-like larvae—with rudimentary thoracic legs and legless abdomen—feed inside the cones of kauri. Upon maturation, the prepupa forms a hardened cell within the cone, wherein it will enter a state of diapause that can extend for a decade or more, to await favorable weather conditions. Prior to emergence, the pupa uses its enormous jaws to free itself from its crypt. Larvae are sometimes so abundant as to be regarded as pests of kauri seeds.

RIGHT | Many archaic moths share similarities with their caddisfly kin; e.g., the wings are held almost vertically along the sides of the abdomen, and the antennae are commonly held forward in repose (neither trait shown in this pinned specimen of *Agathiphaga vitiensis*).

DISTRIBUTION
Northeastern Australia and southwest Pacific Islands (New Caledonia, Solomon Islands, Fiji, and nearby islands)

IMPORTANT GENERA
Agathiphaga

HABITAT
Kauri forest

HOST ASSOCIATIONS
Kauri (*Agathis*)

CHARACTERISTICS
- Wingspans from 0.9–1.1 in (24–27 mm)
- Antennae held forward as in caddisflies
- Pupae with functional jaws; haustellum (proboscis) absent
- Two sets of tibial spurs on both second and third pair of legs
- Larvae with abdominal prolegs; two stemmata (without corneal lens)

HETEROBATHMIOIDEA: HETEROBATHMIIDAE
HETEROBATHMIID MOTHS

This, one of the oldest lineages of Lepidoptera to persist to the present day, is represented by a single genus (*Heterobathmia*) with three described species and at least twice that many undescribed members. Extant *Heterobathmia* are restricted to southern beech (*Nothofagus*) forests of Patagonia. The spring-flying adults are very small, metallic day-fliers with many superficial similarities to Eriocraniidae, their northern hemisphere analogs.

The larvae are leafminers in new spring foliage of southern beech (*Nothofagus*) and make full-depth blotch mines, not unlike those of eriocraniids. They are unusual among leafminers in having the capacity to abandon their mines and form a new mine in a second leaf—although this is surely not commonly done, as the thoracic legs are greatly reduced and the abdominal prolegs are wholly lacking. Prepupal larvae drop to the ground and form a cocoon belowground in which they await the return of spring.

LEFT | In many ways, heterobathmiids are a window to the past, likely preserving many of the same postures, host associations, larval habits, and other traits common to their ancestors. This is *Heterobathmia pseuderiocrania*.

DISTRIBUTION
Patagonian region of South America

IMPORTANT GENERA
Heterobathmia

HABITAT
Mesic forest

HOST ASSOCIATIONS
Southern beech (*Nothofagus*)

CHARACTERISTICS
• Wingspans from 0.3–0.4 in (9–11 mm); FW never with falcate apex; HW broad (nearly as wide as FW), rounded, unpatterned; FW with some metallic gold scaling

• Head sparsely covered with erect, hairlike scales

• Lacking mesotibial spurs

• Larvae with seven stemmata (other moths have only six)

ERIOCRANIOIDEA: ERIOCRANIIDAE
ERIOCRANIID MOTHS

These are splendid little moths and are among the more readily observed lineages of relict Lepidoptera. The 30 or so species are parsed across 5 genera—all are confined to temperate areas of the Holarctic. They are especially well represented across North America, where they have radiated on the rich oak fauna. The adults are active by day and night in the very early spring. While modest in wingspan, they make up for it in beauty—adults are gorgeous creatures, with many scaled in lustrous golds and purples.

The legless larvae are miners in new spring foliage, sometimes maturing before the leaves of their hosts have fully expanded. Indeed, one of the ways eriocraniid mines can be recognized is that the area of the leaf where the egg was inserted often tears as the leaf enlarges, leaving a little hole in the blade. Another clue as to their presence is that the larval feculae are concatenated in strings. Larvae overwinter as prepupae belowground.

TOP RIGHT | Eriocraniids are exquisite creatures when viewed with a lens, bedecked in lustrous purples and golds.

RIGHT | Archaic moths tend to have sparse, hairlike body setae—more modern moths have dense scaling that includes many flattened scales.

DISTRIBUTION
Temperate regions of North America and Eurasia

IMPORTANT GENERA
Eriocrania, *Dyseriocrania*, and *Eriocraniella*

HABITAT
Forests, woodlands, chaparral, and coast scrub

HOST ASSOCIATIONS
Oak (most), birch, and rose families

CHARACTERISTICS
• Wingspans from 0.3–0.5 in (7–12 mm); HW broad (nearly as wide as FW), rounded, unpatterned; FW often with metallic gold and purple scaling

• Head sparsely covered with erect, hairlike scales and possessing short haustellum

• Mesotibia with only single tibial spur (most moths have paired tibial spurs)

• Females with piercing ovipositor

ACANTHOPTEROCTETOIDEA: ACANTHOPTEROCTETIDAE
ARCHAIC SUN MOTHS

This ancient lineage is represented by seven described species. Currently, these are placed in two genera: *Acanthopteroctetes*, from western North America and South Africa, and *Catapterix*, from Eurasia. An undescribed genus is known from the Peruvian Andes. *Acanthopteroctetes* may represent as many as four genera.

The adults are small, undistinguished, and rare in collections and, as a consequence, these relict moths are one of the most poorly studied microlepidopteran families. So far as known, larvae are host-specialized, full-term leafminers that make blotch mines, removing much of the green tissues within. The fully fed larva exits its mine, drops into leaf litter or soil, and spins a cocoon. Winter is passed either as a larva in the mine (*Acanthopteroctetes unifascia*) or as a prepupa on or belowground. Temperate taxa are univoltine.

RIGHT AND INSET | With the exception of the Micropterigidae, most archaic families are believed to be internal feeders as larvae. Life histories are known for only two: both are leafminers. This adult is *Acanthopteroctetes bimaculata*.

DISTRIBUTION
Western North America, Eurasia, South America, and South Africa

IMPORTANT GENERA
Acanthopteroctetes and *Catapterix*

HABITAT
Forests, woodlands, and chaparral

HOST ASSOCIATIONS
Life histories known for just two North American species; one of these on currant (*Ribes*) and one on buckthorn (*Ceanothus*)

CHARACTERISTICS
• Heterogeneous collection of gray, black, and metallic gold moths

• Wingspans from 0.3–0.6 in (7–15 mm)

• Head with erect, hairlike scales

• Frenulum present: as paired bristles in *A. unifascia* and single bristle in *Catapterix*

• Mesotibia with only single tibial spur (most moths have paired tibial spurs)

NEOPSEUSTOIDEA: NEOPSEUSTIDAE
ARCHAIC BELL MOTHS

Archaic bell moths are so divergent in appearance and posture that many entomologists have failed to recognize them as moths. The lineage is represented by just a single family with 4 genera and 14 species: 5 of these are from southern South America and the remainder are endemic to Assam, China, Myanmar, and Taiwan. The adults, somewhat sparsely scaled in various earth tones, can be found active during both day and night.

Very little is known about the biology of these moths, particularly their early stages. Adults of *Apoplania* and *Neopseustis* may be bird-dropping mimics that rest on the upper side of leaves with their wings held flat over the abdomen. The former genus is associated with bamboo forests.

The apex of the female ovipositor bears small teeth, which is suggestive that the female inserts her eggs into plant tissues. Certainly, this is a family greatly in need of study.

LEFT | Unmoth-like in appearance, neopseustids like this *Neopseustis meyricki* are easily overlooked as belonging to "some other insect order." The resting posture and long antennae are diagnostic. Its caterpillar is an external feeder on members of the grape family (Vitaceae).

DISTRIBUTION
Southeast Asia and southern South America

IMPORTANT GENERA
Apoplania and *Neopseustis*

HABITAT
Mesic to wet forests and woodlands; mountains

HOST ASSOCIATIONS
One species known from *Ampelopsis* (grape family)

CHARACTERISTICS
• Wingspans from 0.6–1.1 in (15–27 mm); wings very broad, with prominent veining, often thin and somewhat translucent; held flat over body (most) or in tentiform position

• Antennae long, often exceeding FW length, drawn under wings at rest

• Mandibles present; haustellum scaled, internally with two food canals

HEPIALOIDEA: HEPIALIDAE
GHOST MOTHS

This is among the most diverse and ecologically important lineages of archaic moths, with a history that traces back to the Mesozoic. More than 600 species in 62 genera have been described, with their diversity greatest across the temperate areas of South America, Africa, and Australia. Included are some of the most magnificent moths in size and beauty, with members of *Aenetus*, *Leto*, and *Zelotypia* being striking standouts from the earth-toned adults more typical of the family.

The nonfeeding and short-lived adults can be cryptic in habitat and thus rarely encountered. Much of their scarcity ties to their brevity of activity—many fly for just 20–30 minutes in a given day, usually tied to a brief interval just after sunset. Australian species and others dwelling in drylands tend to fly only with rains, further limiting when they can be observed.

The caterpillars are equally cryptic in habit: most are polyphagous, subterranean root feeders or tunnel into trees. Females broadcast their tiny eggs while in flight—an indication of the polyphagous nature of the larval diets, which in addition to plant roots may include dead leaves, fungi, and soft-bodied insects, as well as their own siblings. One massive Australian species, the Pindi Moth (*Abantiades latipennis*), with a wingspan of

LEFT | Male Ghost Moths (*Hepialus humuli*) fly for only 20 minutes daily, over the latter half of evening twilight and often form small leks. They dissipate the primary sex pheromone, from metatibial androconia while flying in a side-to-side figure-eight loop.

DISTRIBUTION
Cosmopolitan; highest diversity across southern hemisphere and tropics

IMPORTANT GENERA
Aenetus, *Endoclita*, *Fraus*, *Gorgopis*, *Oxycanus*, *Phassus*, and *Thitarodes*

HABITAT
Primarily mesic wooded communities, shrublands, grasslands; also alpine communities

HOST ASSOCIATIONS
Many polyphagous on detritus, fungal hyphae, roots; later instars feeding on roots and commonly boring in aboveground tissues of ferns, gymnosperms, dicots, and monocots; ranging from extremely generalized to host-plant specialists

CHARACTERISTICS
• Wingspans from 1–10 in (25–255 mm)
• Antennae usually characteristically short and beaded; some genera with pectinate

more than 6 in (150 mm), may lay as many as 40,000 eggs.

Hepialoidea possess a unique female reproductive configuration with separate mating and egg-laying openings that are connected by a (caudal) grove or gutter, through which the sperm must swim to reach the spermatheca.

ABOVE | *Abantiades hydrographus*, endemic to Australia, is large-bodied but its eggs are quite small—a single female may produce many thousands of eggs.

ABOVE RIGHT | *Aenetus virescens*, endemic to New Zealand, is the country's largest moth. The larva is a borer in various trees and takes six years or more to reach maturity. The larvae were often eaten by the Māori.

INSET | Hepialid larvae tend to be elongate, muscular, and whitish. The head is often orange, with much-hardened mandibles for feeding on roots or boring into stems.

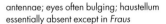

antennae; eyes often bulging; haustellum essentially absent except in *Fraus*

• Abdomen long, often protruding behind wings at rest

• Tentiform resting posture with wings appressed to sides of body

• Larvae elongate, mostly unpigmented, with large, dark pinacula; crochets in several concentric series

MNESARCHAEOIDEA: MNESARCHAEIDAE
MNESARCHAEID MOTHS

This ancient superfamily is represented by a single family endemic to New Zealand. The 14 recognized species are grouped into 2 genera: *Mnesarchaea* and *Mnesarchella*. They are small, rather inconspicuous denizens of mesic forests with lush undergrowth.

The white-and-brown adults are largely diurnal, but some more nocturnal species can be seen at lights. Males account for the majority of specimens in collections as females are more secretive and sedentary in habit. In four species, the female wings are somewhat reduced in size, suggesting they have limited dispersal abilities. Most mnesarchaeids are spring-active.

The mostly unpigmented, elongate caterpillars construct silk-lined tunnels among plants and the litter and duff of the forest floor. In contrast to many relict lepidopterans, the caterpillars are quite mobile, with a full set of thoracic legs and five pairs of abdominal prolegs—the condition common across most modern moth lineages. Larvae feed principally on algae, fern sporangia, fungi, mosses, liverworts, and perhaps forest-floor detritus—all substrates that would have been available in the Mesozoic when the lineage originated. Pupation occurs at the terminus of a tunnel.

LEFT | Mnesarchaeids are small, petite moths like this *Mnesarchaea acuta*. While larval diets in nature are poorly characterized, larvae can be lab-reared on algae, ferns, fungi, and mosses. All these substrates were present in Earth's primeval forests, long before flowering plants evolved; for mnesarchaeids, their old habits seem to have worked well.

DISTRIBUTION
New Zealand

IMPORTANT GENERA
Mnesarchaea and *Mnesarchella*

HABITAT
Wet forest communities

HOST ASSOCIATIONS
Polyphagous on liverworts, ferns, mosses, and fungi, as well as on decaying plant matter of forest floors; evidently angiosperms (flowering plants) are ignored

CHARACTERISTICS
- Wingspans from 0.3–0.5 in (7–12 mm)
- Antennae two-thirds FW length, held upward and to side when perched
- Rudimentary mandibles; short, coiled, functional haustellum
- Sternum of first abdominal segment bilobed
- Larvae elongate and highly mobile with extremely long spinneret

NEPICULOIDEA: NEPTICULIDAE
PETITE LEAFMINING MOTHS

With over 850 described species, Nepticulidae rank as the most evolutionarily successful early diverging lineages of moths. Twenty-two genera are recognized. They are common elements of the moth fauna of all regions and continents where woody plants abound. Fossil (leaf-mine) evidence for the group extends back to the Middle Cretaceous. They are among the smallest moths, with wingspans of only 0.2–0.4 in (4–9 mm). The adults are largely nocturnal.

The family is best known for its great diversity and abundance of its leaf mines. Likely more than 98 percent feed on various angiosperms (flowering plants); a few mine monocots, including graminoids. Virtually all have narrow host ranges. A few induce and feed within galls. Others tunnel in bark, fruits, and meristems. Leaf mines may be serpentine (most) or blotches, but in either case, the mine details are often species specific; for instance, how feculae are deposited within the mine. Notably, some larvae complete their development in fallen leaves, with their mining confined to patches of green tissues that they are able to keep alive.

LEFT | Nepticulids like this *Ectoedemia atricollis* are very tiny. The large "eye cap" at the base of the antenna is made from off-white, flattened scales.

INSET | Larva (visible at mine terminus) of *Stigmella castaneaefoliella*.

DISTRIBUTION
Cosmopolitan

IMPORTANT GENERA
Acalyptris, Ectoedemia, Pectinivalva, Stigmella, and *Trifurcula*

HABITAT
Primarily in any temperate and tropical community with woody plants

HOST ASSOCIATIONS
Extremely diverse; mostly woody angiosperms but also herbaceous perennials and graminoids; diversity greatest on oak in northern hemisphere and myrtaceous hosts in southern hemisphere

CHARACTERISTICS
• Tiny moths with wingspans usually from 0.2–0.3 in (4–8 mm)

• Wings often drawn over abdomen to greater degree than in most archaic lineages

• Tiny spines visible over surface of wings (visible with lens)

• Eyecap of scales at base of antennae

ADELOIDEA: ADELIDAE
FAIRY MOTHS

This charismatic family includes more than 350 species placed into just 5 genera. Adelids are among the most favored of moths for those who have an appreciation for microlepidopterans. They are most diverse in temperate regions—Europe is home to more than 50 species.

The family's hallmark is their remarkably long antennae, with those of *Adela* and *Nemophora* being especially noteworthy. In extreme instances, the antennae may exceed the adult's wingspan by a factor of two or three, and as such put significant drag on their flight. Additionally, the female antennae often sport attractive metallic black or purple scales along the basal portion of the shaft.

Most adelids are diurnal. As is commonly the case among day-active moths, they are frequently rendered in metallic and otherwise splendid colors; dusk- and night-active species have predictably more subdued coloration. Nearly all are spring-active and avid nectarers.

DISTRIBUTION
Cosmopolitan

IMPORTANT GENERA
Adela, *Cauchas*, *Ceromitia*, and *Nemophora*

HABITAT
Temperate and tropical forest openings and edges; woodlands, shrublands, chaparral, serpentine outcrops, deserts, and other open communities

HOST ASSOCIATIONS
Mostly low, not woody, plants (i.e., annual and perennial wildflowers); majority thought to have specialized host associations

The early stages are woefully understood and in great need of study. Females lay their eggs in the developing ovaries of their host plants. The first instar feeds on the developing seeds, then tunnels its way out and drops to the ground, where it spins a case and feeds on dead leaves. Some females are believed to lay their eggs directly into leaf litter. However, so few species have been reared that their preferred diets and larval behaviors remain very much a mystery awaiting study.

FAR LEFT | Fairy Moths like this *Adela reaumurella* are a favorite of those interested in microlepidopterans. Their beautiful colors, conspicuous dancing flights, and local nature make them worthy quarry.

LEFT | *Cauchas*, like this *Cauchas fibulella*, resemble *Adela* but tend to have shorter antennae. Like *Adela*, females insert eggs individually into seeds of the host. Early instars drop to the ground, spin a silken case, and then mature on leaves.

CHARACTERISTICS
- Wingspans usually from 0.2–0.8 in (6–20 mm); wings often with metallic scaling
- Often with exceptionally long antennae, especially in males
- Head with erect, hairlike scales
- Females with piercing ovipositor for inserting eggs into host tissue

TOP | *Nemophora staudingerella*, endemic to eastern Russia and Japan, is a gift of nature for those with the wherewithal to appreciate the little things that run the world. Its life history remains unstudied.

ABOVE | More than 90 species of *Ceromitia* have been described, mostly from Africa and southern South America. As might be guessed, the antennae are so long that they constrain flight.

ADELOIDEA: HELIOZELIDAE
SHIELD-BEARER MOTHS

The 125 species or so of named shield-bearers are grouped into a dozen genera, with the vast majority placed in the nominate genus *Coptodisca*. Likely at least twice this number remain to be described from the tropics and southern hemisphere.

While adult *Coptodisca* are among the smallest moths, the minute adults are exquisitely rendered in metallic steel, silver, and orange, accentuated with bold black and white scaling. The adults are diurnal but seldom encountered. The best way to observe and photograph shield-bearers is to locate their leaf mines and then raise the larvae to adults.

Initially, larvae make a winding linear mine that is abruptly enlarged with all tissues consumed except for the upper and lower leaf surfaces. At the mine terminus, the larvae cut out an oval leaf section from both leaf surfaces and fashion a case (shield). So encased, the prepupal larvae wander away or drop to the ground, then attach with a button of silk to a safe site. The vacated mines with oval cutouts are usually the first tip-off to the presence of these exquisite aerial plankton.

ABOVE | Heliozelid larvae are leafminers. The oval cutouts, fashioned by the prepupal larvae into purse-like "shields," are diagnostic and may remain on a host for months.

LEFT | What shield-bearer moths lack in size, they make up for in beauty. *Antispila* (shown here) have white fascia and spots; *Coptodisca* are ornately scaled in silver, fiery orange, pearly white, and black.

DISTRIBUTION
Cosmopolitan

IMPORTANT GENERA
Antispila, Aspilanta, Coptodisca, and *Heliozela*

HABITAT
Temperate and tropical forests, woodlands, and shrublands

HOST ASSOCIATIONS
Woody plants, especially trees and shrubs; commonly using members of birch, dogwood, grape, heath, myrtle, oak, and rose families

CHARACTERISTICS
• Wingspans mostly from 0.2–0.3 in (5–8 mm) but some to 0.4 in (10 mm); FW oblanceolate and HW lanceolate with long fringe scales; wings often black with metallic scaling

• Head covered with smooth, silvery scales

• Posture and wing position of resting adult diagnostic with anterior elevated

ADELOIDEA: INCURVARIIDAE
LEAFCUTTER MOTHS

The 50 incurvariids are placed in 11 genera, with species and genera yet to be described. What is and what is not an incurvariid is still under inquiry; for example, *Lampronia* is treated here as a prodoxid but in other works as an incurvariid. Diversity is greater in the Old World, with Australia likely to host the greatest number of species.

The larvae are pale and flattened, with reduced prolegs. All are thought to start as leafminers. Some Australian taxa are full-term leafminers, but most northern hemisphere incurvariids vacate the mine and cut out an oval to subquadrate leaf section and then use the excised section to construct a flattened shelter within which it will feed. As the caterpillar matures, it will cut out larger leaf sections and incorporate these into the case—the cutouts and larval feeding damage are highly diagnostic. When fully fed, the prepupa drops to the ground, where it will pass the winter. Alternatively, members of the genus drop from the mine directly to the ground to build their cases. Pupation occurs the following spring. A few have been reported as minor pests.

LEFT | Many leafcutter moths have spots that cross over the dorsum when the adults are at rest, such as this *Incurvaria masculella*. The bristly head, often creamy to orange in color, also helps with recognition. The caterpillars of several species form full-depth blotchmines (inset).

DISTRIBUTION
Cosmopolitan

IMPORTANT GENERA
Incurvaria, Paraclemensia, Perthida, Phylloporia, and *Vespina*

HABITAT
Forests, woodlands, and mountains

HOST ASSOCIATIONS
Specialists on members of many woody plants, including birch, maple, honeysuckle, myrtle, oak, rose, and sugarbush families

CHARACTERISTICS
• Wingspans usually from 0.2–0.5 in (5–12 mm)
• Antennae 0.4–0.8× FW length
• Head set with numerous, erect, hairlike scales
• Male valve with numerous spines and lacking pectinifer
• Apex of ovipositor depressed

ADELOIDEA: PRODOXIDAE
YUCCA MOTHS AND KIN

RIGHT | Female *Tegeticula yuccasella* packing pollen from her maxillary tentacles (elaborated palpi) onto the style of a different flower from which the pollen was collected. Next she will back down the style and insert one or two eggs into the developing ovary.

About 100 species have been described in 9 genera. The family is especially diverse across arid regions of North America, but also occurs in Eurasia, Africa, and South America. More than 70 species have been recorded from the United States. Much of the family's evolutionary success links to their highly specialized association with *Yucca*.

Female *Parategeticula* and *Tegeticula* are the exclusive pollinators of yucca. Immediately before or after pollination, female *Tegeticula* insert one to a few eggs into the developing seed pod. In *Parategeticula*, the eggs are laid in a groove on the flower, and the larvae feed within a mass of galled tissue that is produced as the seed pod develops. The association is obligate—neither the plant nor moths survive without the other. Each of the 40–50 species of yucca has a dedicated *Tegeticula*—17 species have been recognized to

DISTRIBUTION
Mostly North American; also Eurasia and southern Africa

IMPORTANT GENERA
Greya, Lampronia, Parategeticula, Prodoxus, and *Tegeticula*

HABITAT
Deserts and drylands, grasslands, forest edges and woodlands, chaparral, and other open communities

HOST ASSOCIATIONS
Yucca and related genera; saxifrages and umbellifers in *Greya*; currant, rose, and other plant families in *Lampronia*

CHARACTERISTICS
• Wingspans mostly from 0.3–0.7 in (7–18 mm); wings often patterned in gray and white; lacking metallic scaling
• Antennae short, only one-third of FW length

LEFT | Species of *Prodoxus* are freeloaders on *Tegeticula* and *Parategeticula*, the bona fide pollinators of North America's 40 species of yucca. Larvae of all but one *Prodoxus* feed in the floral stalks and other sterile tissues of the inflorescence and thus are fully reliant on successful pollination services carried out by their cousins. Yucca inflorescences that go unpollinated abort, blacken, and wither, taking with them developing *Prodoxus* caterpillars. Shown here are *Prodoxus decipiens* gathered in a yucca flower, waiting for the floral stalk to elongate, into which they will insert their eggs.

this point. An even larger number of prodoxids have evolved that are essentially freeloaders on the services of the pollinating species. The "cheaters" feed on the seeds of yucca (two species of *Tegeticula*), the stalk of the floral spike (most *Prodoxus*), or on other yucca tissues (a few *Prodoxus*).

All prodoxid larvae feed internally. Most drop from the host as prepupae to form a cell in soil. *Prodoxus* form pupal cells at the terminus of the larval mine within their *Yucca* hosts. Desert taxa have an extraordinary capacity to tolerate years of drought before emerging. One cohort of *Prodoxus y-inversus* yielded adults over a 30-year period.

- Head covered with erect scales
- Females with long, laterally compressed ovipositor for inserting eggs into host tissue
- Bursa with two stellate signa

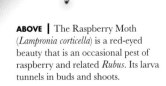

ABOVE | The Raspberry Moth (*Lampronia corticella*) is a red-eyed beauty that is an occasional pest of raspberry and related *Rubus*. Its larva tunnels in buds and shoots.

PALAEPHATOIDEA: PALAEPHATIDAE
PALAEPHATID MOTHS

This relict southern hemisphere lineage includes some 57 extant species distributed across 7 genera. Additional species await discovery and description, especially from South America.

Males have a rich array of secondary sex scales (androconia); these are secreted in pockets and folds on the wings, as well as on the pleural regions of the abdomen. Presumably the androconia play important roles in pheromone release during courtship displays that precede copulation. In other moths, such scales are especially important where nocturnally active congeners fly together, with the species-specific pheromones released from the androconia helping to sort out species identities and prevent mispairings.

Life histories are known for only a few palaephatids. So far as known, early instars start out as leafminers and then exit the mines to form leaf shelters in which they complete their development. Like other leafminers, the larval diets are specialized. Late instars are fully legged with well-developed thoracic legs and five pairs of abdominal prolegs. The pupal shell is extruded from the cocoon prior to emergence.

LEFT | Like many relict lineages, palaephatids have a Gondwanan distribution, being restricted to temperate areas of the southern continents, and represent survivors of an ancient lineage that may have been substantially more diverse and ecologically important in epochs past. Note the bristly head and long, filament-like antennae.

DISTRIBUTION
Australia including Tasmania, South Africa, and South America

IMPORTANT GENERA
Azaleodes, Metaphatus, Palaephatus, Ptyssoptera, and *Sesommata*

HABITAT
Woodland and shrubland communities

HOST ASSOCIATIONS
Mostly on Proteaceae; others using Verbenaceae (*Azaleodes*)

CHARACTERISTICS
- Wingspans mostly from 0.3–0.6 in (7–16 mm); wings typically brown to gold, often lustrous, with or without white patches
- Antennae with two rows of bidentate scales along dorsal side
- Head vertex with erect, hairlike scales; dorsum of thorax often with raised scale tuft
- Very short haustellum (proboscis); labial palpi curved slightly upward

TISCHERIOIDEA: TISCHERIIDAE
TRUMPET LEAFMINER MOTHS

LEFT | The resting posture, with the anterior elevated and legs positioned like stilts beneath the body, is diagnostic for the family. Tischeriid caterpillars (inset) are leafminers: the body is dorsoventrally compressed and attenuates rearward.

Just over 100 species have been described in just 3 genera, with many awaiting description, especially from across Mexico. Nearly half the named species have been described from the United States. Trumpet leafminers defy the trends in most plants and animals in that their species diversity is highest in middle latitudes where oaks abound. Most are multiple-brooded, with abundance increasing through second (summer) and third (fall) generations.

Their common name derives from the larval mine, which tends to get broader in each instar; in some, the terminus of the mine flares outward like the horn of a trumpet. Many are edge-miners that use silk to draw the leaf margin over the top of the mine. As in many leafminers, the mine shape and position on the leaf are typically species-specific—and frequently easier to identify than the rather undistinguished adults.

The larvae keep a tidy abode, chewing slits in the floor of the mine through which they release their excreta. Pupation occurs in the mine. All overwinter as pupae. Adults are common at light; many can be net-collected at dusk.

DISTRIBUTION
Cosmopolitan except for Australia

IMPORTANT GENERA
Astrotischeria, *Coptotriche*, and *Tischeria*

HABITAT
Forest edges and openings, woodlands, shrublands, chaparral, grasslands, and meadows

HOST ASSOCIATIONS
Majority on oaks and related genera; also members of aster, buckthorn, mallow, and rose families

CHARACTERISTICS
• Wingspans often less than 0.3 in (9 mm) but some to 0.4 in (11 mm); wings often uniformly straw, yellow, or gray

• Male antennae with numerous long cilia

• Head with distinctive erect scales over vertex; smooth-scaled front, triangular in frontal view

• Posture of resting adult diagnostic with anterior elevated (as shown above)

DITRYSIAN MICROLEPIDOPTERA

Ditrysians include moth superfamilies possessing two female reproductive openings: one for oviposition and a second opening used solely for copulation. Sperm swim from the bursa copulatrix, through a narrow duct, the ductus seminalis, to the spermatheca (see page 51). As the eggs are moving down the common oviduct, the spermatheca will release sperm into the oviduct. Over 98 percent of described species of Lepidoptera are ditrysians, with this figure set to rise as most of the moths that have yet to be named are ditrysians. Why this disparity should be so great may never be known, but the difference could be rooted to the evolution of a second reproductive opening and its associated structures. At the anterior end of the second reproductive opening is a voluminous pocket, the bursa copulatrix, which accommodates the spermatophore from male partners. This enlarged pouch has made it possible for males to transfer nutritionally important resources, in addition to sperm, during copulations: commonly proteins, carbohydrates, lipids and fatty acids, sterols, and sundry defensive compounds that can protect their mates and can even be passed forward to the offspring (see pages 50 and 225).

This section (up to page 163) describes the ditrysian microlepidopterans, with more than 20 superfamilies. All but two of these are profiled: Simaethistoidea and Papilionoidea. The former consists of just six species, including two species of Whalleyanidae. The latter, the butterflies, are treated in a separate volume in this series. Note that Tineoidea appears twice in the phylogeny, as what is currently being treated in the Tineoidea includes at least two separate lineages, which serves as a reminder of the nascent and dynamic state of the classification of microlepidopterans. Mimallonoidea share

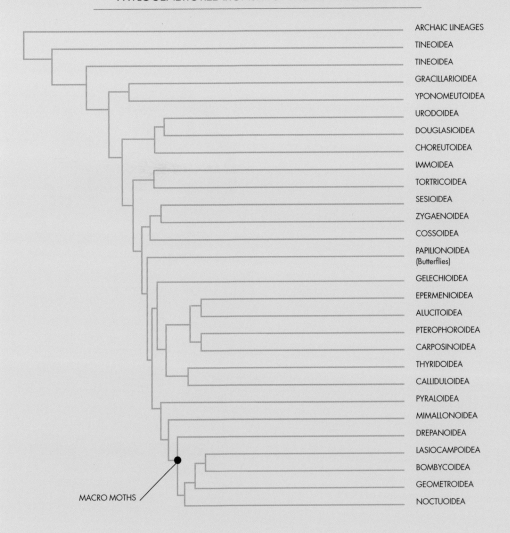

characters with both micro- and macrolepidopterans; they are included here as microlepidopterans as the larvae are shelter formers and their crochet arrangement is typical of micros.

Ditrysian microlepidopterans are exceedingly diverse. They lack the characteristics for macrolepidopterans reviewed on page 164. The caterpillars mostly feed internally or from shelters. Their crochets are arranged in a complete circle, and in most the proleg is elongate (longer than thick). (When poked many dart backwards with alacrity.) The study of microlepidopterans offers galaxies of opportunities.

TINEOIDEA: PSYCHIDAE
BAGWORM MOTHS

This is a large, cosmopolitan family with over 1,350 species and 240 genera; more than 80 percent of the species diversity is confined to the Old World. The family is exceedingly diverse in form—a dozen subfamilies are recognized—making characterization of the adults nearly impossible; many are difficult to differentiate from clothes moths (Tineoidea).

Adults may be either diurnal or nocturnal. Males of day-flying taxa are often beelike, with clear wings, and furtive; some are blisteringly fast when in flight. Females can be fully winged, brachypterous, wingless, or larviform. The latter may even lack functional legs and other appendages—having traded their investments in flight and walking for elevated reproductive output. Those that have forfeited on flight remain concealed in the larval case. Males have an extensible abdomen that they insert into the female's case during mating. Females commonly lay their eggs within the larval case and die.

Neonate caterpillars often "balloon"; that is, drop on threads of silk to be carried away on winds. The larvae are broadly polyphagous, consuming live and even dead plant matter. A few are predaceous on small insects and will use silk to secure the carcasses of their victims onto their

RIGHT | Bagworm moth cases are infinitely diverse in detail yet often have species-specific features that allow accurate identification of the occupant, long after they have been vacated. Indeed, most of the cases I have encountered proved to be abandoned.

DISTRIBUTION
Cosmopolitan

IMPORTANT GENERA
Acanthopsyche, Apterona, Arrhenophanes, Dahlica, Narycia, Oiketicus, Oiketicoides, Psyche, Thyridopteryx, and *Typhonia*

HABITAT
Generalized in ecology, inhabiting deserts and badlands to savannas, scrublands, and tropical rainforests

HOST ASSOCIATIONS
Most broadly polyphagous and perhaps better thought of as omnivorous; some predaceous

CHARACTERISTICS
• Wingspans from 0.2–1.7 in (5–44 mm)

• Male antennae often serrate, pectinate, plumose, or otherwise elaborated; simple in females

• Mouthparts, including palpi, greatly reduced in both sexes

cases. The size, shape, and nature of what plant matter and other objects are woven into the case wall are often species specific. A few are pests of ornamental and orchard crops. Several psychids are parthenogenetic. The pupa of the Fangalabola moth (*Deborrea malgassa*) is harvested and eaten by indigenous peoples in Madagascar.

INSET | While larvae of *Oiketicus* bagworms can be so abundant as to become pests, adults are seldom encountered. The males are diurnal, nonfeeding creatures with only one mission—mating; the females are larviform that stay ensconced in their larval case.

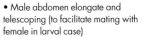

- Male abdomen elongate and telescoping (to facilitate mating with female in larval case)
- Larvae forming a thick-walled silken case to which leaf fragments and other debris are attached

ABOVE | Male of *Canephora hirsuta* freshly eclosed from its case as well as its pupal shell. Note its elaborate antennae, which it will soon employ to seek out a calling female.

TINEOIDEA: TINEIDAE
FUNGUS MOTHS

This is a large, diverse, and taxonomically complex family with over 2,400 described species placed into about 360 genera distributed worldwide. The higher classification is in great flux: Currently 14 different subfamilies are recognized, with some of these treated as valid families in some works; many genera have yet to be reliably placed into any subfamily.

Tineids are exceedingly diverse both morphologically and biologically, with fascinating life histories that extend into the bizarre. A few are live-bearing with the eggs hatching within the female's abdomen.

Tineid larvae typically feed on fungi, lichens, dead leaves, and other detritus, and in so doing are an important family of decomposers and nutrient recyclers. The caterpillars typically construct a silken tube spun in the larval substrate; some form cases, as in the common clothes moths that infest woolens. The caterpillars are elongate, often whitish, with strongly differentiated prothoracic shield and pinacula.

Tineids are exceptional among animals in being able to digest keratin—the protein that makes up hair, nails, fur, feathers, turtle shells, and horns. They have capitalized on this novel adaptation to

RIGHT | The hundreds of *Acrolophus* species, most of which remain undescribed, are believed to feed on the roots of grasses and other plants, as well as on nonliving organic matter. This is *Acrolophus popeanella*.

DISTRIBUTION
Cosmopolitan

IMPORTANT GENERA
Acrolophus, Amydria, Erechthias, Infurcitinea, Monopis, Nemapogon, Opogona, Tinea, and *Tineola*

HABITAT
Drylands to tropical rainforests; diverse in moist ecosystems with abundant fungi and detritus

HOST ASSOCIATIONS
Exceedingly diverse; most feed on fungi and lichens; many detritivores eating dead leaves and wood (with fungi), guano and scat, feathers, fur, and wool

CHARACTERISTICS
• Wingspans from 0.2–2.0 in (5–50 mm)
• Wings held in tentiform position; adults often make diagnostic scuttling runs
• Head often with distinctive tuft of densely packed, erect, hairlike scales

flourish as inquilines in bird, reptile, and mammal nests, and as decomposers that feed on the corpses of various vertebrates.

Beyond these exceptional niches, tineids include species that inhabit the nests of social wasps and termites; a few that consume the remains that accumulate below spiderwebs; others that feed on scat and guano; and several that invade our homes to feed on tapestries and clothes made from wool.

- Haustellum absent or short and uncoiled, with two halves often loosely associated; maxillary palpus often five-segmented (or absent); labial palpus frequently with stiff lateral bristles
- Females with telescoping ovipositor

TOP AND INSET | Clothes moths (*Tinea*) are among the most notorious pests of stored woolens and ornamental textiles in homes. In the inset, two larvae can be seen moving their cases to new locations.

ABOVE | Like many tineids, the biologies of *Monopis*, such as this *Monopis dorsistrigella*, are poorly known. At least three species have been reared from bird nests, where they feed on nest detritus.

GRACILLARIOIDEA: GRACILLARIIDAE
LEAF BLOTCH MINER MOTHS

This is a splendid and aged lineage of cosmopolitan moths, with forewing patterns that rival those of the most beautiful Lepidoptera. However, a microscope or other means of magnification will be needed to appreciate their beauty, as they are small. Almost 2,000 species have been described in about 100 genera, but perhaps three times this number await discovery.

Gracillariid hindwings are essentially aerial oars composed of long fringe scales that allow them to paddle through the air on calm nights.

The vast majority mine leaves as larvae, but other plant tissues are targeted. A few induce galls

BELOW | *Caloptilia* (shown here) and kindred genera are splendid moths with a diagnostic tripod stance.

DISTRIBUTION
Cosmopolitan, including islands

IMPORTANT GENERA
Acrocercops, Caloptilia, Cameraria, Marmara, Parornix, Phyllocnistis, and *Phyllonorycter*

HABITAT
Forests, woodlands, shrublands, and chaparral, as well as many open, sunny habitats

HOST ASSOCIATIONS
Mostly woody, broad-leaved plants; many hosted by members of the birch, oak, rose, and willow families; oak may host richest fauna globally

CHARACTERISTICS
• Wingspans mostly from 0.2–0.6 in (5–15 mm); wings with very long fringe scales; usually with costal strigulae; FW often brightly colored in red, orange, or white; HW tan to gray, unpatterned

in which they will mature. Two subfamilies, Phyllocnistinae and Lithocolletinae, feed full-term within their tunnels and pupate in their mines. The nominate subfamily, Gracillariinae, are diverse in habit, often feeding as a leafminer in early instars and then exiting their mines to form a leaf shelter, within which they will complete their larval development. These and many lithocolletines have hypermetamorphic development, meaning they have two radically different larval forms: a flattened, legless, nearly blind mining form that molts into a cylindrical, legged, fully eyed caterpillar that feeds within its leaf shelter or blister mine.

Gracillariids have narrow host ranges—such that getting a host-plant determination will greatly facilitate moth identification. The mines, shelters, galls, and other signs of the larval habits are often diagnostic—essentially providing a semipermanent record of the larvae's feeding behaviors. Because of this, leafminers are increasingly employed in ecological studies.

- Antennae without eyecap; flagellum narrowly filiform with one row of encircling scales per flagomere, 0.8–1.8× FW length
- Head often with erect tuft scales that spread outward; smooth, triangular front (face) extending well below eyes
- First two to five instars legless; most with fourth to final instars legged but missing prolegs on A3; *Marmara* and some other full-term miners legless through last instar

ABOVE | *Phyllonorycter* is an enormous, cosmopolitan genus with over 400 described species. Most form bubble-like mines on leaf undersides of woody plants. Shown here is *Phyllonorycter leucographella*.

INSET | By holding leaf mines up to a light source (such as the sun), it is often possible to see if a mine is occupied with a larva or parasitoids or is vacant. Shown here is a vacant mine of *Parectopa lespedezaefoliella*.

GRACILLARIOIDEA: BUCCULATRICIDAE
RIBBED COCOON-MAKER MOTHS

Nearly all of the 300 species are members of the nominate genus *Bucculatrix*. Depending on the authority, one to three additional genera are recognized. Recent molecular data are suggestive that the family will be reassigned to the Yponomeutoidea. Much of the species diversity is endemic to North America, where dozens of additional species await description.

Larvae are host-specialized leafminers (most) or gall-makers. Most species make a short, full-depth mine in early instars, and then exit the mine in a middle instar to feed externally, skeletonizing small patches of leaf tissue. A few are full-term miners—these often carve out blotch mines. One clade forms stem galls in composites. The prepupa of *Bucculatrix* fashion a distinctive ribbed cocoon.

Much of the evolutionary success of the family is tied to oaks (Fagaceae) and members of the sunflower family (Asteraceae)—two plant families that are especially diverse in North America.

LEFT | Raised forewing scales, tuft of erect head scales, and triangular face are diagnostic of these very small moths like this *Bucculatrix coronatella*.

INSET | The minute, ribbed cocoons of bucculatricids are diagnostic—these are typically spun below the host on understory vegetation or forest-floor substrates.

DISTRIBUTION
Principally North America; small numbers across Eurasia, Africa, South America, and Australia

IMPORTANT GENERA
Bucculatrix and *Ogmograptis*

HABITAT
Forests, shrublands, chaparral, grasslands, and deserts; favoring open, sunny habitats

HOST ASSOCIATIONS
Aster, birch, and oak families; others on members of buckthorn and rose families

CHARACTERISTICS
• Wingspans from 0.2–0.6 in (5–15 mm)
• Antennae with large eyecap
• Head often with erect tuft of scales that spread outward; smooth, triangle front (face) extending well below eyes
• First two instars legless; third to final instars fully legged

YPONOMEUTOIDEA: YPONOMEUTIDAE
ERMINE MOTHS

The 9 to 11 families that make up the Yponomeutoidea are one of the most heterogeneous and ill-defined assemblages of Lepidoptera. Likewise, the nominate family, the Yponomeutidae, is exceptionally diverse in morphology and biology. No characters are known that unite the members of the superfamily or family, which suggests that both taxa are unnatural and will be divided and otherwise reorganized. The 370 described species of Yponomeutidae have been placed into 95 genera, parsed across 6 subfamilies. While the family is found worldwide, most of the diversity is in temperate areas of both hemispheres.

Keeping in step, the larvae are also diverse in form and habit. Many feed externally within a webbing of silk spun over the host. While ermine moth caterpillars are generally solitary, members of at least two subfamilies spin communal webs. A few European *Yponomeuta* are eruptive and can cover trees, cars, and whole neighborhoods in ghostly sheets of silk. Leaf-mining is widespread; other yponomeutids are internal feeders in buds, fruits, shoots, and stems.

ABOVE | Members of the nominate genus, *Yponomeuta*, have eruptive population cycles, with extraordinary outbreaks where host trees are stripped of leaves and wandering caterpillars enshroud nearby objects with sheets of silk.

INSET | The gregarious caterpillars of *Yponomeuta* form a communal nest that is enlarged through each instar. Even within the safety of their shroud, the larvae aggregate.

DISTRIBUTION
Cosmopolitan

IMPORTANT GENERA
Swammerdamia, *Yponomeuta*, and *Zelleria*

HABITAT
Most associated with forests, woodlands, and shrublands

HOST ASSOCIATIONS
Exceedingly diverse, including conifers; most are host-plant specialists on woody plants

CHARACTERISTICS
- Wingspans from 0.2–0.7 in (6–18 mm)
- A8 often with pleural lobes that extend along sides of genital capsule
- Haustellum well developed and unscaled
- Larvae with crochets often in two or more series; often feeding under cover of loose webbing of silk

YPONOMEUTOIDEA: HELIODINIDAE
SUN MOTHS

LEFT | Sun moths or heliodinids, such as this *Aetole bella*, are often spectacularly rendered in metallic silver, orange, and red scales, with silver scales frequently raised to make a set of spots presumably to yield a jumping spider look.

This small family includes some spectacular moths— if they were larger, they would surely rank among the most favored of insects. Conservative estimates for the family richness hover around 70 species and 12 genera; but different authors recognize as many as 400 species and 60 genera.

The larvae are diverse in habit: tunneling into flowers, buds, leaves, and seeds, often under a filmy sheet of silk, as is common among yponomeutoids. The adults of many genera are thought to mimic small jumping spiders, with forewings that have closely situated, raised metallic scales, which resemble a set of jumping spider eyes, and silvery striae across the wings, which approximate legs. Some make jerky movements and short hops, in a fashion that resembles the movement of jumping spiders. The diurnal adults make good quarry for the moth photographer.

DISTRIBUTION
Widespread, but principally North and South America

IMPORTANT GENERA
Aetole, Embola, Epicroesa, Heliodines, and *Lithariapteryx*

HABITAT
Deserts, grasslands, and tropics

HOST ASSOCIATIONS
Specialized on many dicots; important hosts include four o'clock, evening primrose, fig-marigold, goosefoot, and ginseng families

CHARACTERISTICS
• Wingspans from 0.2–0.6 in (5–16 mm); wings commonly in reds and oranges with silvery striae; many metallic

• Antennae thickened

• Head smooth-scaled

• Dorsal side of hind tibiae smooth-scaled

• Pupae often angulate; pupation occurs in mine or under wispy silk cover

GELECHIOIDEA: XYLORYCTIDAE
TIMBER OR HERMIT MOTHS

The 500-plus species fall into approximately 60 genera. The family is well nested within the Gelechioidea, although its relationship to other family-level gelechioid taxa is unclear. Most of the species and generic diversity is endemic to Australia, where the family is among the more ecologically important groups of microlepidopterans. Timber moths abound in myrtaceous (eucalyptus) and acacia forests in eastern Australia.

The larvae are diverse in habit. Many tunnel into flowers, bark, twigs, branches, and stems. While most are specialized in diet, others are generalized, especially those that tunnel into wood.

The caterpillars commonly form shelters. Several feed on callous tissue about the tunnel entrance, while others drag leaf fragments to their tunnels and then silk these about the entrance area, so that these may be later consumed. In either case, the tunnel entries are covered with a webbing into which frass has been interwoven. A few are case-makers that carry their shelters with them. A modest number are significant pests of orchard and ornamental trees.

LEFT | Timber moths are stout microlepidopterans with somewhat squarish forewings, visible here in *Maroga melanostigma*.

INSET | Resting postures are often diagnostic for moths. Xyloryctids wrap the wings about the abdomen, with the forelegs held forward, as seen here in *Cryptophasa rubescens*.

DISTRIBUTION
Sub-Saharan Africa, Indo-Australia, and Polynesia

IMPORTANT GENERA
Cryptophasa, Lichenaula, Thyrocopa, and *Xylorycta*

HABITAT
Woodlands and forests

HOST ASSOCIATIONS
Proteaceae and Myrtaceae commonly, but including many families of woody plants

CHARACTERISTICS
• Wingspans mostly from 0.4–2.7 in (11–70 mm); FW often squarish with angulate tornus; HW broad, sometimes wider than FW

• Antennae usually bipectinate in males

• Abdominal segments 2–6 with spines across caudal portion of each plate (terga)

• Labial palpi typically large, upcurved

• Larval abdominal segments 1–8 with ring around SD1; secondary SV setae on A3–7

GELECHIOIDEA: OECOPHORIDAE
OECOPHORID MOTHS

Currently, more than 3,300 species and 313 genera are recognized, although the taxonomic limits of the family are in flux. The family is closely allied to both the Pterolonchidae and Peleopididae.

Oecophorids are hyperdiverse in Australia, where more than 2,300 species and 200 genera have been recognized. In many eucalyptus and other myrtaceous woodlands, oecophorids are among the most numerous moths at lights.

The adults are highly divergent in size and form, with no character uniquely shared by its current members—which is often an indication that the group is not monophyletic, that is, contains taxa that need to be reclassified elsewhere.

Larvae commonly construct leaf shelters on living leaves, but an even larger fraction feed on nonliving plant tissues: leaf litter, bark, and fallen logs. They are the primary decomposers of eucalyptus leaf litter across much of Australia. *Philobota* larvae live in vertical tunnels in the soil and emerge at night to forage on grasses and herbs; a few species are pasture and cereal crops pests.

Members of *Trisyntopa* are inquilines in parrot nests, where they feed principally on droppings. Many Australian oecophorids are case-makers.

LEFT | Oecophorid moths, though small, can be handsomely patterned, particularly the diurnal species of the nominate subfamily, represented here by *Oecophora bractella*. Members of the genus *Fabiola* are exquisite jumping spider mimics.

DISTRIBUTION
Cosmopolitan, but with two-thirds of world fauna in Australia

IMPORTANT GENERA
Barea, Borkhausenia, Eulechria, Inga, Philobota, Promalactis, and *Pleurota*

HABITAT
Woodlands and forests; fewer numbers in heathlands and tropical communities

HOST ASSOCIATIONS
Living and dead leaves; greater diversity evidently on decaying plant matter; many feed on eucalyptus and other Myrtaceae

CHARACTERISTICS
• Wingspans mostly from 0.2–1.2 in (5–30 mm); some genera folding wings flat over dorsum

• Abdominal segments 2–6 without spines across caudal portion of each plate (terga)

• Labial palpi long, typically with slender, upcurved terminal (third) segment

• Gnathos often fused to tegumen

GELECHIOIDEA: DEPRESSARIIDAE
FLAT-BODIED MOTHS

What is and what is not a Depressariidae varies among works, museums, and taxonomic treatments. The taxon is most often accorded its own family status or subsumed within the Elachistidae. Many works include a number of gelechioid subfamilies within the Depressariidae. Clearly, the higher classification of the Gelechioidea is greatly in need of a modern, comprehensive assessment.

Depending on which treatment is examined, estimates of species-level diversity range upward to 3,500 species and beyond. Here, only two subfamilies are featured: Depressariinae and Stenomatinae.

Depressariines include more than 600 species and 80 genera. Their larvae feed on buds, flowers, seeds, and stems; the majority tie up and feed within leaf shelters. Nearly all are host-plant specialists. Depressariines have radiated on host plants that manufacture furanocoumarins, reactive cyclic ring compounds that in the presence of ultraviolet light (sunlight) cross-link DNA—dooming many unprepared caterpillars and other herbivores. By fashioning their leaf shelters to exclude light and feeding at night, depressariine caterpillars are able to circumvent the toxic effects of these phytotoxins.

Adults of many species overwinter under bark and other concealed sites. A few are important pests.

ABOVE | Depressariids tend to be rendered in tans, browns, grays, and other earthen tones. The wings are folded flat over the dorsum, which is unusual for microlepidopterans.

INSET | The labial palpi of gelechioids are distinctly upcurved and the tongue scaled at its base.

DISTRIBUTION
Cosmopolitan, but most diversity in temperate areas

IMPORTANT GENERA
Agonopterix, *Depressaria*, *Eutorna*, *Exaeretia*, and *Psorosticha*

HABITAT
Forests to early succession habitats

HOST ASSOCIATIONS
Exceedingly diverse; common hosts include members of carrot, citrus, legume, mallow, and rose families

CHARACTERISTICS
• Wingspans mostly from 0.4–1.0 in (9–24 mm); FW and HW broad, folded flat over dorsum

• Gnathos of male genitalia spined

• Caterpillars with three L setae anterior to the prothoracic spiracle; secondary setae usually absent or restricted to anal prolegs

GELECHIOIDEA: DEPRESSARIIDAE: STENOMATINAE
STENOMATINE MOTHS

Stenomatines are treated here as a subfamily of Depressariidae, although some place them as a subfamily of Elachistidae. In general appearance, they have much in common with the Depressariinae (page 115). Regardless of classification, stenomatines warrant their own account because of their diversity and ecological importance in the Neotropics.

Over 1,200 species have so far been named. They are among the most commonly encountered lepidopteran caterpillars across the jungles of Central and South America. They also boast the largest gelechioids, with some *Timocratica* reaching wingspans exceeding 3 in (80 mm).

Unusually for microlepidopterans, the larvae often feed on mature, fully hardened leaves. Most are leaf-tiers: many make leaf shelters between overlapping leaves. Others tunnel into bark and stems and, like xyloryctids (page 113), include species that feed on callous tissues about the entrance of their tunnels. The larvae tend to be patterned with a large head that facilitates feeding on tough, older leaves.

LEFT | Stenomatines are exceedingly diverse in the Neotropics and account for more than half of all Depressariidae. Several genera mimic bird droppings at rest, as exemplified here by *Antaeotricha schlaegeri*.

DISTRIBUTION
Mostly Neotropical; modest representation in Nearctic and across Old World

IMPORTANT GENERA
Antaeotricha, Cerconota, Chlamydastis, Stenoma, and *Timocratica*

HABITAT
Tropical, premontane, and temperate forests and woodlands; thorn scrub and shrublands

HOST ASSOCIATIONS
Myrtle family (Myrtaceae) hosts close to one third of those with known life histories; Nearctic fauna restricted to oaks

CHARACTERISTICS
• Wingspans from 0.3–3.2 in (8–80 mm); FW broad, squarish; HW rounded and often broader than FW; HW with Rs vein bent toward subcostal vein at end of discal cell

• Antennae often with long, cilia-like setae

• Valve of male genitalia with some apically lobed setae (except *Agriophara*)

• Larvae often modestly flattened

GELECHIOIDEA: COLEOPHORIDAE
CASE-BEARING MOTHS

The boundaries of the family are contested, with conservative treatments recognizing about 1,400 species and 5 genera, with most of these in the nominate genus. More than 1,260 species of *Coleophora* have been described, with many dozens yet to be named, especially from across Mexico and the western United States.

Larvae are leafminers or seed feeders that fashion a silken case that often has species-specific attributes that allow identification. Host-plant identity will also aid identification efforts, as most species have specialized diets. Cases are endlessly varied: most an elongate cone; others cigar-shaped, pistol-shaped, or baggy; and many include leaf fragments. The larval case is enlarged as the caterpillar matures. Coleophorid larvae leave telltale feeding damage: typically there is a single entry hole where the caterpillar has tunneled into the leaf or seed.

Leaf-mining taxa typically consume all the tissues between the two leaf surfaces, leaving a translucent blotch, free of any feculae (which are released through the distal end of the case).

ABOVE | At rest, case-bearing moths such as this *Coleophora albicosta* are narrowly elongate and hold their somewhat thickened, often ringed, antennae forward. Adults tend to be unmarked white to earthen colored; many have lines or streaks that run parallel to the body axis.

INSET | Coleophorid larval cases are intriguingly varied in shape across species and allow instant identification of many species. Shown here is *Coleophora alnifoliae*.

DISTRIBUTION
Cosmopolitan, but most diversity across northern hemisphere

IMPORTANT GENERA
Augasma, *Coleophora*, and *Goniodoma*

HABITAT
Forests to grasslands and deserts

HOST ASSOCIATIONS
Mostly broad-leaved plants, including many forbs; seed heads of graminoids; much Nearctic diversity on composites

CHARACTERISTICS
• Wingspans mostly from 0.3–1.0 in (7–25 mm); wings elongate and narrow

• Antennae thick, ringed, often thickly scaled at base; often held forward in repose

• Commonly with paired, dorsal scale patches over abdominal segments

• Caterpillars case-making with reduced prolegs; crochets in two transverse rows

GELECHIOIDEA: MOMPHIDAE
MOMPHID MOTHS

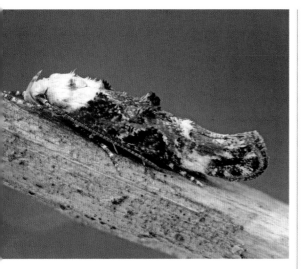

Based on their global diversity of just 7 genera and little more than 120 described species, this gelechioid family would be unworthy of inclusion save for their great beauty, which no doubt ties, at least in part, to widespread diurnality across the family. Momphidae are allied to the Blastobasidae, Scythrididae, and Stathmopodidae.

Larvae are bud, flower, fruit, stem, and root borers, as well as leafminers. Host associations are narrowly specialized, as are the (separate) larval feeding niches when more than a single species share a common host plant.

The family is most diverse in the Nearctic, where much of their species diversity is tied to the evening primrose family. The Neotropics host small radiations on members of the melastome and coffee families. Temperate taxa typically overwinter as pupae, but at least a few overwinter as adults and prepupal larvae.

LEFT | This small family is a favorite among micromoth aficionados: many are quite strikingly colored. The patches of raised forewing scales are diagnostic; a few, with metallic scaling, appear to be jumping spider mimics. Above left: a bird-dropping mimic (*Mompha propinquella*). Left: a flashy diurnally active species (*Mompha raschkiella*).

DISTRIBUTION
Mostly Nearctic, Neotropics, and Palearctic

IMPORTANT GENERA
Mompha, Moriloma, Palaeoamystella, and *Zapyrasta*

HABITAT
Various plant communities, from tropical and temperate forests to prairies and deserts

HOST ASSOCIATIONS
Many use evening primrose principally across Holarctic; also cliffrose, melastome, coffee, and other plant families

CHARACTERISTICS
• Wingspans mostly from 0.3–0.6 in (7–15 mm); wings elongate, moderately narrow, often with raised scale patches; anal veins absent from HW

• Gnathos absent in males; paired signa, each with projecting apodeme, in females

• Caterpillars with two L setae anterior of prothoracic spiracle

GELECHIOIDEA: COSMOPTERIGIDAE
COSMET MOTHS

This is a heterogeneous assemblage of more than 1,800 species and 135 genera of small and often exquisitely patterned animals. While the nominate genus and related core lineages readily define the cosmet moths as deserving their own family, they contain many anomalous genera—expect some genera to be classified elsewhere. Among their ranks are scores of handsome moths that are favorites among those who have interests that include microlepidopterans.

The larvae are exceedingly diverse in habit, feeding in buds, flowers, seeds, fruits, leaves, bark, and roots. The nominate genus and many others are leafminers. Several form leaf, stem, and root galls. A few genera are scavengers; there are even lineages that specialize on aquatic plants. Predation on armored scale insects can be added as still another niche for these perplexingly diverse moths. Host-plant associations are often quite specialized and encompass a wide taxonomic range of plants.

ABOVE | The nominate genus *Cosmopterix* are delicate, fetching creatures, adorned with silvery spots and red to orange scales.

INSET | As caterpillars, some *Cosmopterix* make irregular blotch mines with blind extensions. Shown here is *Cosmopterix zieglerella*.

DISTRIBUTION
Cosmopolitan; particularly diverse in Hawaii (*Hyposmocoma*)

IMPORTANT GENERA
Ascalenia, Asymphorodes, Cosmopterix, Hyposmocoma, Labdia, Limnaecia, Macrobathra, and *Stilbosis*

HABITAT
Many plant communities, from tropical rainforests to deserts, and from aquatic and riparian plant communities to drylands

HOST ASSOCIATIONS
Diverse, with more than 30 plant families, from ferns and gymnosperms to an array of dicots and monocots, including grasses

CHARACTERISTICS
• Wingspans from 0.2–0.9 in (5–23 mm); FW and HW narrow; FW frequently with tufts of raised scales

• Female retinaculum consisting of scale group arising anterior to CuA vein

• Pupae compact; sometimes dorsoventrally flattened

GELECHIOIDEA: GELECHIIDAE
TWIRLER MOTHS

The Gelechioidea contains 21 families and more than 19,000 described species, and at least this many unnamed species. The nominate family includes more than 4,700 species and 510 genera. In many plant communities, gelechiids could be the most diverse microlepidopteran family, but they are so understudied it will be decades before we will know their richness with any precision. For most areas of the planet, estimates will have to come from molecular surveys of species-level genetic diversity as there are far too few gelechiid taxonomists to have any hope of describing most of the world's fauna. The adults range from mundane to beautiful insects, well worthy of greater attention.

The larvae feed on a nearly boundless array of niches, consuming buds, flowers, fruits, leaves, stems, and roots. The caterpillars are concealed

DISTRIBUTION
Cosmopolitan, including oceanic islands

IMPORTANT GENERA
Ardozyga, Aristotelia, Chionodes, Compsolechia, Dichomeris, Gelechia, Gnorimoschema, Helcystogramma, and *Thiotricha*

HABITAT
Diverse in tropical and temperate forests; chaparral and other shrublands; prairies, savannas, and deserts

HOST ASSOCIATIONS
Exceedingly diverse; more than 90 different plant families used, including mosses, conifers, and graminoids

CHARACTERISTICS
• Wingspans mostly from 0.3–0.8 in (7–20 mm) but some to 1.2 in (30 mm); FW narrow to somewhat quadrangular; HW usually falcate with outer margin curving inward below apex

Ditrysian Microlepidoptera

feeders, not visible on the outside of their food plant: most are shelter-formers, leafminers, or gall-makers. Host-plant associations and larval niches tend to be highly specialized. The caterpillars are often boldly marked with spots and/or striping, and as a consequence are more readily identifiable than most microlepidopteran caterpillars.

Gelechiids include many important pests of field crops; the Pink Bollworm (*Pectinophora gossypiella*) is a major pest of cotton worldwide. Several leaf-mining species are important silvicultural pests of conifers.

- Smooth-scaled head; base of haustellum with scales; third segment of labial palpus, slender, upcurved
- Hind tibiae with long scales above

FAR LEFT | This is an enormous family, perhaps with only a third of its species described. Adults range from those that are earthen colored (most) to brightly colored diurnal moths with intricate markings.

ABOVE | Dichomeridines—a subfamily of Gelechiidae with nearly a thousand species—have a distinctive snout with enlarged labial palpi that are directed well forward.

INSET | The Pink Bollworm (*Pectinophora gossypiella*) is an internal borer in developing cotton bolls (flowers) that costs global markets hundreds of millions of dollars each year. Genetically modified cotton varieties are being developed.

ALUCITOIDEA: ALUCITIDAE
MANY-PLUMED MOTHS

This small taxon, of unmistakable identity, is a favorite among moth aficionados. The 216 species and 9 genera are distributed worldwide.

Larvae tunnel into buds, flowers, fruits, and stems; some are gall-formers. Host associations are specialized. *Alucita coffeina* is a pest of coffee across Africa. Temperate species of *Alucita* may overwinter as adults—and are often noted in outbuildings on warmer winter and early spring days.

ABOVE LEFT | The deeply divided forewings and hindwings immediately distinguish members of this family, represented here by *Alucita phricodes*.

LEFT | *Alucita phanerarcha*, from southwestern Africa, is a gaudy relative to most members of alucids, which typically are rendered in grays, browns, and other earthen tones.

DISTRIBUTION
Cosmopolitan, including Oceania; much species diversity in Africa

IMPORTANT GENERA
Alucita, Hexeretmis, Microschismus, Pterotopteryx, and *Triscaedecia*

HABITAT
Mostly forests and woodlands

HOST ASSOCIATIONS
Diverse, commonly utilized hosts include members of the bignonia, honeysuckle, and coffee families

CHARACTERISTICS
- Wingspans mostly from 0.4–0.7 in (11–18 mm); FW and HW usually deeply divided into 6–12 plumes; *Hexeretmis* with merely lobed wing margins; wings held to side at rest
- Smooth-scaled head with raised dorsolateral tufts; hind tibiae sometimes with long, stiff setae above
- Haustellum well developed

PTEROPHOROIDEA: PTEROPHORIDAE
PLUME MOTHS

Pterophorids can be recognized by their exceptionally long, spidery legs and wings that are folded and rolled together into a bundle and held away from the body. The 90 genera include more than 1,400 species.

The adults are commonly seen at lights, although a more sporting way to observe plume moths is to walk open habitats at dusk. They are avid nectarers that can be reliably sought at flowers from nightfall onward. Meadows, rich in floral diversity, are a good place to look for both adults and their caterpillars.

The externally feeding, stubby caterpillars, with their bristly secondary setae, are distinctive; those that bore internally are more grub-like and difficult to find. Most feed on flowers or leaves of annual or perennial herbaceous plants of open habitats.

The Artichoke Plume Moth (*Platyptilia carduidactyla*) is an important pest species.

RIGHT | Plume moths, with their divided wings, are popular subjects among moth photographers as they are often conspicuous when perched. Getting species-level names is not so easy, especially beyond the faunas of Europe and Japan. This is the White Plume Moth (*Pterophorus pentadactyla*), a common European pterophorid.

DISTRIBUTION
Cosmopolitan

IMPORTANT GENERA
Agdistis, Hellinsia, Oidaematophorus, Platyptilia, Pterophorus, and *Stenoptilia*

HABITAT
Favoring early successional and other open, sunny habitats, from wet and mesic meadows to grasslands and deserts

HOST ASSOCIATIONS
More than 25 plant families; most on herbaceous dicots

CHARACTERISTICS
• Wingspans mostly from 0.3–1.2 in (9–30 mm); FW and HW usually deeply divided; resting posture with FW and HW folded and rolled together and held to side; wings not plumed in *Agdistis* and *Ochyrotica*

• Long-legged with metatibia more than twice length of metafemur

• Larvae with secondary setae that are sometimes forked, clubbed, or secretory

• Pupae often conspicuously setose; usually exposed as chrysalis without cocoon

CARPOSINOIDEA: CARPOSINIDAE
FRUITWORM MOTHS

The Carposinoidea includes two relatively small families: Carposinidae and Copromorphidae. The former includes approximately 20 genera and 300 species.

Carposinid larvae are most commonly internal feeders in flowers, fruits, shoots, bark, and stems; others form leaf shelters and galls. Those that bore into stems sometimes feed primarily on the wound-reaction tissues that form about their tunnel entrances.

As is commonly the case for internal feeders, host associations tend to be specialized. Carposinid caterpillars are typically pale and undifferentiated with respect to patterning. A few are fruit pests. The Peach Fruit Moth (*Carposina sasakii*) is a major pest of apple, peach, and pear; the Australian Guava Moth (*Coscinoptycha improbana*) damages citrus, guava, macadamia, peach, pear, plum, and other commercial species.

LEFT | Carposinid adults often have somewhat triangular forewings, which are laid against the substrate at rest, and a pronounced snout.

INSET | *Carposina scirrhosella*, shown here, and relatives are an occasional pest of peaches and related pomes; others are inquilines in stem galls.

DISTRIBUTION
Cosmopolitan; diverse through Australia, New Guinea, New Zealand, and Hawaii

IMPORTANT GENERA
Bondia, *Carposina*, and *Paramorpha*

HABITAT
Woodlands, shrublands, and chaparral

HOST ASSOCIATIONS
More than 20 dicot families recorded: e.g., aster, citrus, heath, macadamia nut, myrtle, oak, and rose families

CHARACTERISTICS
- Wingspans from 0.4–1.6 in (10–40 mm); FW and HW rounded; FW often with raised scale tufts
- Labial palpi upcurved; third segment projecting forward (forming snout)
- Two L group setae on shared pinaculum anterior to prothoracic spiracle; spiracle on A8 enlarged and shifted above level of preceding spiracles

DOUGLASIOIDEA: DOUGLASIIDAE
DOUGLAS MOTHS

This is a small superfamily with just a single family of about 30 species parsed across 3 genera.

The anterior end of the body is elevated when perched. The pale caterpillars are stout internal borers in leaves, petioles, stems, and inflorescences, often with reduced prolegs lacking crochets and very long setae. Pupation occurs at the mine terminus; the pupa is extruded from the host at eclosion.

The habits and life histories of the family are poorly known, especially outside of Europe, and would make for a good study subject for those fortunate enough to discover a colony of these diminutive insects.

ABOVE RIGHT | Small, slaty to gray microlepidopterans that are seldom encountered. *Tinagma ocnerostomella* feeds internally in the pithy stems of bugloss.

RIGHT | The forewings of *Tinagma* often have areas of dark scaling alternating with paler gray sections. Many douglasiids are diurnal and can be seen perching on their favored host by day.

DISTRIBUTION
Mostly temperate regions of northern hemisphere; one Australian species

IMPORTANT GENERA
Klimeschia and *Tinagma*

HABITAT
Woodlands, forest edges and openings, chaparral, and coastal scrub

HOST ASSOCIATIONS
Poorly known but including members of borage, mint, and rose families

CHARACTERISTICS
• Wingspans from 0.2–0.6 in (6–15 mm); HW much narrower than FW; FW gray or brown, somewhat lustrous

• Antennae two-thirds FW length with row of slender scales encircling each flagellomere

• Head smooth-scaled above and conspicuous ocelli

• Larvae stout, pale, with well-defined prothoracic and anal plates; three L group setae on prothorax; paired SV setae on thorax

SCHRECKENSTEINIOIDEA: SCHRECKENSTEINIIDAE
BRISTLE-LEGGED MOTHS

Schreckensteinioidea is a small superfamily with just 4 genera and 12 species. When perched, the adults are recognizable, as they rest with their long-spined hind legs held upward over the body.

The family's common name is a bit of a misnomer in that some Heliodinidae (page 112) and a few oecophorids (page 114) have considerably more bristly hind legs and also hold their hind legs above the body when perched. Presumably the raised legs serve to thwart the attacks of small predators that might try to pounce on the moths.

The larvae feed externally on leaves, removing the green tissues from a feeding site, leaving a network of veining (thus skeletonizing the leaf). All are host-plant specialists. The most well-known species, the Blackberry Skeletonizer (*Schreckensteinia festaliella*), is a Palearctic species that was inadvertently introduced to North America, then later purposefully released into Hawaii to control invasive brambles (*Rubus*).

ABOVE AND INSET | In repose, bristle-legged moths hold their rear legs, bearing two pairs of enormous spurs, over their body. The Blackberry Skeletonizer (*Schreckensteinia festaliella*) is a common resident across Europe, and parts of North America. Its cocoon is unusual in being more of a latticework through which the pupa is fully visible (*inset*).

DISTRIBUTION
Nearly worldwide but not yet known from Africa and Indo-Australia

IMPORTANT GENERA
Corsocasis, *Ptilosticha*, and *Schreckensteinia*

HABITAT
Tropical and temperate forests, woodlands, and wide range of open plant communities

HOST ASSOCIATIONS
Sumac, plantain, and rose families

CHARACTERISTICS
• Wingspans from 0.3–0.5 in (8–12 mm); FW and HW narrow

• Legs long, smooth-scaled, with exceptionally long tibial spurs; hind legs held above body in repose

• Labial palpi upcurved and project forward

• Larvae with long setae; some dorsal setae spatula-like; slender prolegs with few crochets

• Distinctive, exposed, fishnet-like cocoon often spun near feeding site

EPERMENIOIDEA: EPERMENIIDAE
FRINGE-TUFTED MOTHS

Epermenioidea contain only a single family, a dozen genera, and fewer than 100 described species. The phylogenetic relationships of epermeniids with other moths are still under study—perplexingly, they share similarities with Copromorphoidea, Schreckensteinioidea, and Yponomeutoidea. Life histories are only known for a fraction of the global fauna.

As caterpillars, many tunnel into fruits and especially seeds; several are leafminers; others feed under a sparse silken web or form leaf shelters; and a few are gall-formers. Many are gregarious as larvae. Host associations are specialized, with members of the carrot family hosting about half those with known life histories across the northern hemisphere. The Garden Lance-wing (*Epermenia chaerophyllella*) is a minor garden and crop pest of plants in the carrot family.

LEFT AND INSET | Fringe-tufted moths, usually rendered in gray or brown, possess raised scale tufts along the lower (inner) margin of the forewing—the first tuft is roughly twice as large as the more distal. Garden Lance-wing (*Epermenia chaerophyllella*) shown here feeds on the leaves of many members of the carrot family across Europe and Asia Minor.

DISTRIBUTION
Cosmopolitan; well represented across Indo-Australia

IMPORTANT GENERA
Epermenia, *Gnathifera*, and *Ochromolopis*

HABITAT
Forests, woodlands, meadows, many open habitats; denizens of mesic communities

HOST ASSOCIATIONS
Most commonly members of carrot family; also bittersweet, coffee, mistletoe, pea, pine, sandalwood, and spurge families

CHARACTERISTICS
• Wingspans from 0.3–0.8 in (8–20 mm); FW and HW narrow; FW fringe scales also including flattened scales

• Adults with anterior end of body elevated at rest; often with raised scale tufts along midline (conspicuous in lateral view)

• Hind tibiae with abundant short, stiff setae (bristles)

• Often in open-mesh cocoon

URODOIDEA: URODIDAE
FALSE BURNET MOTHS

RIGHT AND INSET | Urodids have a bullet-like profile at rest. *Urodus parvula*, shown here, purportedly ranges from New York to Brazil, but such ranges are best verified with molecular data. The caterpillar is an occasional pest of bay and avocado.

This small family consists of 8 genera and 80 species, with highest diversity in the New World tropics. Urodidae and the recently described Ustyurtiidae from Kazakhstan comprise the known fauna of the superfamily.

Life histories are known for only a modest fraction of the world fauna. The Bumelia Webworm Moth (*Urodus parvula*) is evidently widely polyphagous, although most urodids are believed to be host-plant specialists. Larvae spin an open-meshed cocoon that is suspended on a silken petiole.

That the caterpillars are often conspicuously colored with oranges and yellows, and the pupa is essentially fully visibly through the net-walled cocoon, are suggestive that both the larva and pupa will prove to be distasteful.

DISTRIBUTION
Mostly tropical, with much diversity in Central and South America and Asia

IMPORTANT GENERA
Spiladarcha and *Urodus*

HABITAT
Forests and woodlands

HOST ASSOCIATIONS
Poorly known

CHARACTERISTICS
• Wingspans from 0.4–1.5 in (11–37 mm); distal half of discal cell bent downward; upper surface of FW with erect scale tufts; males of *Urodus* with scent scales along upper edge of HW

• Hind tibiae with long, hairlike scales

• Labial palpus with scale tuft on underside of second segment

• Larvae with elongate, slender prolegs constricted at midlength

IMMOIDEA: IMMIDAE
IMMID MOTHS

This pantropic family of microlepidopterans consists of about 250 species spread across 10 genera. Species diversity is highest in Southeast Asia, with only modest representation in Australia, Africa, and the Neotropics. More than half the species have been described in the nominate genus, *Imma*.

The adults are variable in appearance: many resemble oecophorids; others with broader wings are choreutid-like in aspect and posture. At rest, many hold the first pair of legs forward, with the anterior of the body slightly elevated. Diurnal species are often handsomely marked with various oranges but many are tan or gray and unheralded—these also tend to be the taxa with more acute forewing apices.

The elongate caterpillars of *Imma* are often green initially, with long, conspicuous setae (with microscopic barbs) and conspicuously tapering abdominal segments; their anal prolegs are directed backward. The last instars are often dark with conspicuous setae issuing from prominent warts (verrucae). The externally feeding caterpillars have an animated thrashing response when threatened and are quick to drop on a belay line that they can ascend after danger has passed.

ABOVE | *Imma*, an enormous genus with over 150 species, is most diverse across Australia, Indonesia, and Southeast Asia. Many are brown to black with an orange, yellow, or white marking that cuts into or extends across the forewing.

INSET | The upcurved labial palpi and resting posture of the adults help with recognition. At rest the wings look squared off. Many raise the antennae upward like antlers.

DISTRIBUTION
Pantropical; very diverse in Southeast Asia; poorly represented in Neotropics

IMPORTANT GENERA
Moca and *Imma*

HABITAT
Forests

HOST ASSOCIATIONS
Poorly known; believed to be specialists on woody plants, including conifers

CHARACTERISTICS
• Wingspans mostly from 0.6–1.4 in (15–35 mm); wings somewhat elongate with rounded tornus; FW apex rounded in many but frequently angulate; HW broader than FW

• Haustellum unscaled; labial palpi upcurved covering face with second segment densely scaled

• Males often with courtship brushes at terminus of abdomen

• Larvae elongate, tapering rearward; slender prolegs with crochets aligned in linear series; prothorax with three L setae

CHOREUTOIDEA: CHOREUTIDAE
METALMARK MOTHS

BELOW | *Brenthia* adults mimic jumping spiders both in appearance and habit, making herky-jerky movements while holding the wings upward to enhance their deceptive ploy. Shown here is *Brenthia pavonacella*.

Choreutids, the sole representative family in Choreutoidea, include some of the most beautifully and behaviorally fascinating moths. The 400-plus species are parsed across a couple of dozen genera, with much of their diversity tied to the tropics. Molecular data suggest Choreutoidea and Urodoidea are sister taxa.

DISTRIBUTION
Cosmopolitan, with diversity highest in tropics

IMPORTANT GENERA
Brenthia, Choreutis, Prochoreutis, and *Tebenna*

HABITAT
Forests and woodlands to grasslands and deserts; temperate species often in prairies, meadows, and other open plant communities

HOST ASSOCIATIONS
Greatly varied; members of fig family support rich fauna across New and Old World tropics; composites support many *Tebenna*; other important hosts include the borage, dipterocarp, mint, pea, and rose families

CHARACTERISTICS
• Wingspans from 0.2–0.8 in (5–20 mm); wings broad, many with metallic markings; FW roughly triangular; *Brenthia* hold

130   Ditrysian Microlepidoptera

Adults are mainly diurnal, especially in temperate regions; tropical faunas include both diurnal and nocturnal lineages and some species that appear to be active both day and night.

Brenthia, with over 80 described species, are especially compelling jumping spider (Salticidae) mimics, with silver striae on the wings that resemble legs and metallic scale patches that resemble a set of eyes. Adults will perch on the upper side of leaves and dart forward in short dashes, not unlike salticid movements. The ruse is so good that salticids that are equal to or smaller in size to the moths will often engage in a territorial display or scurry off when approached by a *Brenthia* adult.

Choreutid larvae are host-plant specialists. Many feed beneath amorphous rafts of silk that prevent the ingress of some would-be predators. *Brenthia* caterpillars fashion an escape hole in the leaf, below their silken webs—if any portion of the silk lattice is displaced, the caterpillar will be instantly alerted and dart through its escape hole, where it will hold until the uninvited visitor has departed. A few are occasional orchard pests.

their wings to the side to mimic jumping spiders

• Antennae filiform; males often with cilia along lower surface and/or flattened scales along upper surface

• Haustellum scaled

• Larvae with elongate, slender prolegs; L1 and L2 setae on A1–A8 widely separated

TOP | *Hemerophila* and kindred choreutids have radiated on figs (Moraceae). *Hemerophila diva* is a creature of extraordinary beauty.

INSET | Choreutid caterpillars have exceedingly long setae that detect any movement in the silk webbing that the larvae spin over their feeding site—an early warning system for trouble.

BOTTOM | *Tortyra slossonia* and many other choreutids, including *H. diva*, feed on figs—how they are able to circumvent the milky latex sap of *Ficus*, a potent deterrent to most caterpillars, while feeding is worthy of more study.

GALACTICOIDEA: GALACTICIDAE
GALACTICID MOTHS

Galacticids are a very small and understudied group of microlepidopterans. The world fauna consists of just three genera and fewer than two dozen species. The phylogenetic placement of the superfamily relative to other Lepidoptera is unresolved; they are only provisionally placed here between the Immoidea and Tortricoidea.

The most familiar representative is the Mimosa Webworm (*Homadaula anisocentra*), which is a common pest of mimosa and honey locust (*Gleditsia*), especially where the insect has been introduced (for example, in the United States).

The larvae feed gregariously within filmy layers of silk, skeletonizing the included leaves; damage worsens through successive summer and fall generations. Galacticids overwinter in a dense silken cocoon, sometimes spun in bark crevices.

BELOW AND INSET | Adult and larva of Mimosa Webworm (*Homadaula anisocentra*). This native of East Asia was inadvertently introduced into the eastern United States, where it has become an occasional pest of various mimosas and honey locust.

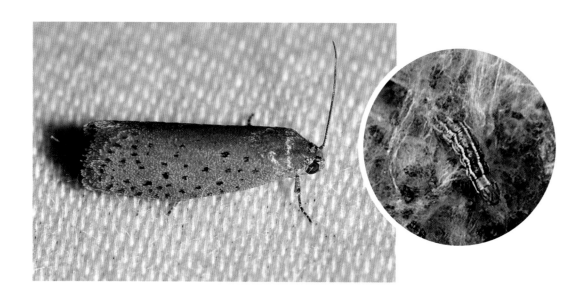

DISTRIBUTION
Endemic to Old World; introduced into North America

IMPORTANT GENERA
Homadaula and *Tanaoctena*

HABITAT
Forests and woodlands

HOST ASSOCIATIONS
Pea (Fabaceae) family

CHARACTERISTICS
• Wingspans from 0.3–0.7 in (8–17 mm); wings rounded, relatively broad, subquadrangular

• Antennae filiform or pectinate, often with basal pecten that covers part of eye at rest

• Tergum of A8 extends rearward, enveloping genitalia

• Female mating pore (ostium bursae) on end of short tube projecting from intersegmental membrane between A7 and A8

• Pronotal shield of larvae including L setae

TORTRICOIDEA: TORTRICIDAE
LEAFROLLER MOTHS

Tortricidae are among the most diverse and ecologically successful moths, with more than 11,400 described species representing more than 1,170 genera. As with most hyperdiverse taxa, there is no single character that allows for their recognition, although features listed below will diagnose the vast majority of genera.

Tortricids are especially abundant in temperate woodlands and forests, where their caterpillars are a mainstay for birds and other insectivores, and as such are likely a major driver of the timing of migratory movements and the nesting success of songbirds. The caterpillars are concealed feeders: most feed in tied leaf clusters or roll a leaf edge to form a shelter; others are borers; a few mine thick leaves and bark; a number of olethreutines are gall-forming; and members of at least one tribe, especially in Australia, feed on fallen leaf litter.

Host associations tend to be specialized among Olethreutinae and Chlidanotinae, while many Tortricinae are polyphagous. The family includes many important pests.

TOP AND LEFT | *Acleris* can be wonderfully and bewilderingly polymorphic—treatises could be written about the wildly different forms of *A. variana*, shown here, *A. hastiana*, and others. Adults of *Acleris* overwinter under bark and in wood sheds, log piles, and other refugia. During warm winter nights they can be drawn to fermenting baits of the moth hunter.

FAR LEFT | *Hilarographa* are tortricids of great beauty but often go undetected because they are largely diurnal, evidently do not visit flowers, and are inconspicuous in habit.

DISTRIBUTION
Cosmopolitan, but especially diverse in temperate regions

IMPORTANT GENERA
Refer to subfamily accounts

HABITAT
Most plant habitats from equator to poles

HOST ASSOCIATIONS
Exceptionally diversified, with trees and shrubs hosting most species; many stem and root borers on small shrubs and perennials

CHARACTERISTICS
- Wingspans mostly from 0.3–1.4 in (7–35 mm) but some over 2 in (5 cm); FW often subquadrangular; HW usually broader than FW, often with concave outer margin
- Antennae filiform, only extending to about half FW length
- Head rough-scaled; snouted profile; labial palpi with apical segment shortened

TORTRICOIDEA: TORTRICIDAE: TORTRICINAE
TORTRICINE LEAFROLLER MOTHS

Tortricines are among the world's most abundant and diverse moths. They are especially well represented in the southern hemisphere. Their higher classification is in a state of renaissance now that molecular data are becoming available to untangle their evolutionary history—expect many taxonomic changes in coming years across the 11 currently recognized tribes.

Tortricines are well represented across the temperate zones of both hemispheres, although the Atteriini, many Cochylini, and others are richly represented in tropical regions.

The caterpillars are shelter-formers and include tip-tiers, leaf-rollers, and a few leafminers; others are internal feeders in meristems, stems, and roots, as well as in fruits and seeds. Borers are especially diverse in herbaceous perennials and other low-growing plants. Members of the tribes Epitymbiini and some Sparganothini feed primarily on leaf litter. Tortricines include many orchard and forest pests; several budworms (*Choristoneura*) cause massive defoliations to spruce and other conifers across the northern hemisphere.

LEFT AND INSET | The Spruce Budworm (*Choristoneura fumiferana*) and its congeners are among the most destructive pests of conifer forests in North America, with outbreaks sometimes causing major timber losses. The caterpillars feed on many conifers but are especially damaging to fir and spruce.

DISTRIBUTION
Cosmopolitan; high diversity across many temperate regions

IMPORTANT GENERA
Acleris, Archips, Argyrotaenia, Choristoneura, Cnephasia, Cochylis, Homona, Platynota, and *Sparganothis*

HABITAT
Diverse, from subarctic regions to equator; especially wherever woody plants dominate, but especially temperate forests, woodlands, and shrublands

HOST ASSOCIATIONS
Diverse, largely reflecting species richness and ecological dominance of woody plants; both dietary specialists and generalists

CHARACTERISTICS
• Small to medium-small moths; FW often squared off

• Antennae usually with two rows of scales along flagellum

• Aedeagus and juxta articulating (not fused)

• Cubital pecten usually absent

TORTRICOIDEA: TORTRICIDAE: OLETHREUTINAE
OLETHREUTINE LEAFROLLER MOTHS

Olethreutinae are ecologically, evolutionarily, and economically among the most important microlepidopterans. About 5,000 species have been described, currently parsed across 5 tribes.

Male olethreutines display an extraordinary array of different male (androconial) scent scales that are deployed during courtship. These scales, laden with pheromones, tend to be most prevalent and ultrastructurally elaborate when closely related species feed on the same host plant, and as such have an elevated risk of making mating mistakes.

The caterpillars of most olethreutines form silken shelters, tying and rolling leaves together; others are borers in shoots, fruits, stems, and roots; and a few form galls. The largest tribe, Eucosmini (1,600 species), are mostly borers, especially in composites (Asteraceae). Olethreutines include many serious orchard and crop pests: important examples include Codling Moth (*Cydia pomonella*) and Oriental Fruit Moth (*Grapholita molesta*).

RIGHT | For every tortricine there may be two olethreutines, such as this Codling Moth (*Cydia pomonella*).

INSET | Jumping Bean Moth (*Cydia saltitans*) caterpillar. Once matured, it is able to move the hollowed-out bean about by arching its head back and then slamming it forward into the opposite wall of the seed.

DISTRIBUTION
Cosmopolitan; especially diverse in northern hemisphere; not especially well represented across tropical regions

IMPORTANT GENERA
Bactra, Cryptophlebia, Cydia, Dichrorampha, Epinotia, Grapholita, Olethreutes, Pammene, and *Polochrista*

HABITAT
Virtually all plant communities where woody plants dominate; especially temperate forests, woodlands, and shrublands

HOST ASSOCIATIONS
Diverse, largely reflecting species richness and ecological dominance of woody plants

CHARACTERISTICS
• Small to medium-small moths; at rest appearing somewhat bullet-shaped from above, with wings enveloping sides of body

• Antennae usually with one row of scales along flagellum

• Aedeagus and juxta fused

• Cubital pecten usually present

SESIOIDEA: SESIIDAE
CLEARWING MOTHS

BELOW | Sesiids are extraordinary moths in coloration, biology, and habit. All mimic wasps, as exemplified here by *Pennisetia marginata*, and virtually all are diurnal. Adults nectar at flowers but do so furtively and tend to be difficult to approach.

More than 1,500 species of these perennial favorites among moth enthusiasts have been recognized; many more still await description. Species diversity of these fascinating wasp mimics peaks in the arid regions of the temperate zone. So far as known, tropical diversity is modest.

DISTRIBUTION
Cosmopolitan, especially in temperate regions of both hemispheres

IMPORTANT GENERA
Bembecia, Carmenta, Melittia, Nokona, Sophona, and *Synanthedon*

HABITAT
Wide variety of habitats, but especially deserts, grasslands, steppes to above timberline; well represented in open habitats with low plants and flowers for nectaring

HOST ASSOCIATIONS
Nearly all are host-plant specialists on herbaceous and woody perennials, with much of their diversity tied to low-growing plants

CHARACTERISTICS
• Wasplike in coloration and behavior; wingspans mostly from 0.4–2.4 in (10–60 mm); narrow, elongate wings, often with clear areas free of scales

• Trailing edge of FW recurved downward and leading edge of HW recurved upward such that wings interlock into single airfoil

The adults, Batesian mimics of wasps, are among the most beautiful and interesting moths, but little is known of their adult habits as most are diurnal, wary, evasive, and easily overlooked as being wasps. Their mimicry goes well beyond their wing coloration, as their legs, posture, and details of flight have all been honed by natural selection to augment their resemblance to stinging wasps. Most have a well-developed haustellum and are avid flower visitors; a few are nonfeeding with a reduced tongue.

The caterpillars tunnel in twigs, stems, and roots and are especially prevalent in ground-level root–stem tissues. Some specialize on large seeds (mostly in the tropics). A number form galls, and a few are predaceous on scale insects. Sesiids include important pests of fruit trees and ornamental plants, and several are crop pests; for example, the Squash Vine Borer (*Eichlinia cucurbitae*).

The most reliable way to survey for the adults, or at least the males, is to purchase and deploy synthetic pheromones that include chemical constituents (or close analogs) of the sex pheromones released by virgin females. Where hosts are known, their larvae, when collected with ample host tissue, can be secured and reared to the adult stage.

- Antennae thickened and relatively short; often clavate; frequently pectinate or bipectinate in males
- Legs handsomely clothed in long scales that enhance wasp mimicry
- Larvae elongate, mostly unpigmented, except for melanized, shiny orange-brown head and darkened prothoracic and anal shields; short prolegs with two transverse bands of crochets

TOP | *Alcathoe caudata* is an amazing creature and should be on every moth hunter's must-see list—but be forewarned: it is not easy quarry. The male abdomen bears a striking orange "tail" set with hundreds of orange scales.

ABOVE | More than a hundred species of *Carmenta* are native to the New World. Unlike most moths, sesiids are more diverse in temperate ecosystems than those of tropical realms.

COSSOIDEA: COSSIDAE
GOAT MOTHS

INSET | Carpenterworm Moth larva *Prionoxystus robiniae* lab-reared from eggs on potato and carrots. In the wild, larvae feed on the wood of various trees and may take 1–4 years to mature.

Cossids are among the most massive of all moths and give pause regarding the use of the term "microlepidopterans." The 1,000 species of this family are best represented in the Neotropics, Africa, East Asia, and Indo-Australia.

The nonfeeding adults are short-lived and seemingly focused entirely on reproduction. They are exceptionally strong fliers—males in particular seem virtually impossible to see, net, or photograph unless they have been attracted to a light or are

DISTRIBUTION
Cosmopolitan; diverse in Neotropics, Africa, East Asia, and Indo-Australia

IMPORTANT GENERA
Azygophleps, Cossulus, Dyspessa, Endoxyla, Givira, Phragmataecia, and *Trismelasmos*

HABITAT
Temperate and tropical forests, woodlands, and chaparral and other shrublands, with high diversity in deserts, savannas, and other drylands

HOST ASSOCIATIONS
Diverse in nature; mostly trees and shrubs; root feeders often feed on a variety of plants, including grasses

CHARACTERISTICS
- Wingspans mostly from 0.5–9 in (12–230 mm)
- Large, muscular, robust-bodied, with long abdomen that usually extends beyond wings at rest

being baited to a calling virgin female. Females, often much more massive than the males, have enormous abdomens, which are packed with hundreds of eggs.

The caterpillars are borers in live or dead wood, although some of the subterranean species feed externally on root tissues. While many are dietary generalists, most are thought to have specialized host associations.

Eggs are typically laid in bark crevices of the host; a few lay large rafts of eggs. Larval development may extend over more than a year. Several are pests of timber and orchard trees. The larvae of several large species in the genus *Endoxyla*, known as witchetty or witjuti grubs, are eaten by indigenous peoples of Australia.

The Maguey Worm (*Comadia redtenbacheri*) is an abundant borer in the crown of agaves. As a marketing ploy, some Mexican mescal distilleries add a single caterpillar to bottles of mescal as an indicator of authenticity. In many regions of Mexico, the "worms" are added to tacos and other Mexican dishes.

OPPOSITE | The Wood Leopard Moth (*Zeuzera pyrina*), native to Europe and Asia but now established in eastern North America, is a general feeder and pest, especially of fruit trees.

BELOW | Reed Leopard (*Phragmataecia castaneae*), native to Eurasia and northern Africa. Cossids have a distinct habitus: they are elongate, with the abdomen extending beyond the wings, possess a robust thorax, and bear short, pectinate antennae, with the apex often narrowed.

• Antennae usually less than half FW length (some reaching three-quarters), thick, often with long rami in males

• Abdominal terminus often extensible, allowing females to insert eggs into bark crevices

• Larvae large, grub-like; strongly melanized prothoracic plate; L1 and L2 sharing common pinaculum on A1–A8

COSSOIDEA: CASTNIIDAE
BUTTERFLY OR SUN MOTHS

These marvelous moths, comprised of about 115 species, are a favorite among collectors. Nearly all are diurnal with attractively colored hindwings, and are frequently mistaken for butterflies. Their knobbed antennae, similar to true butterflies, add to the confusion.

The adults are wary and fast fliers—many more will be seen than captured. The territorial males often dart from perch to perch, with the head positioned so they have a good view of local flyways. When perched on a stick, the wings are often held below horizontal and the anterior end of the moth elevated.

Contrary to most moths, some castniid males produce the primary sex pheromone, which they use to mark and stake out their territories.

As is typically the case for cossoid moths, the larvae are borers. Larval development may extend over more than a year. A few are pests of banana, palm, and sugarcane.

ABOVE LEFT | Castniids are fabulous moths—large, brightly colored, diurnal—renowned for their wariness and exceptionally strong flight: startle one, and likely it will be the last time you see the creature. Shown here is *Synemon jcaria*.

LEFT | Butterfly moths often have brightly colored hindwings. In *Athis inca* (shown here) the "secondaries" are orange with a contrasting black border. Castniids, like butterflies, have knobbed antennae.

DISTRIBUTION
Diverse in Neotropics; modestly represented in Southeast Asia and Australia; one introduced species in Europe

IMPORTANT GENERA
Athis, Castnia, Synemon, and *Tascina*

HABITAT
Tropical and subtropical forests

HOST ASSOCIATIONS
Monocots, including arrowroots, bananas, bromeliads, grasses, orchids, and palms

CHARACTERISTICS
• Robust, powerful, day-flying moths with wingspans from 1.0–7.5 in (25–190 mm); in repose, broadly triangular FW covers brightly colored HW

• Females with telescoping abdominal terminus for insertion of eggs

• Larvae large cylindrical borers in monocots; head retracted into prothorax; strongly sclerotized prothoracic plate (as in Cossidae)

ZYGAENOIDEA: EPIPYROPIDAE
PLANTHOPPER PARASITE MOTHS

This, one of the 12 families in the Zygaenoidea, warrants inclusion in this work not for its diversity, beauty, or size but instead for the sheer novelty of its life history: the larvae are parasitoids on the bodies of planthoppers (Heteroptera: Fulgoroidea) primarily, but also cicadas and related auchenorrhynchans. The 32 described species of epipyropids fall into 9 genera distributed worldwide.

The adults are small, clothed in flat black, brown-black, or gray scales and rather nondescript. The females are hyperfecund, laying hundreds to thousands of tiny eggs on the host plant of their targeted planthopper.

The minute first instars are capable crawlers, with much of the body made of muscle needed to drive their cursorial thoracic legs. Once attached to a host, the caterpillar silks itself to the body and molts to a pale, featureless form. As it grows, it covers itself with flakes of flocculent wax, much of which will be incorporated into the wall of its cocoon, spun on vegetation below the host plant of the planthopper.

LEFT | Epipyropids are extraordinary moths. Three of the grub-like caterpillars are visible here feeding on the hemolymph of the cicada host.

INSET | Adult epipyropids tend to be small, broad-winged, and darkly scaled.

DISTRIBUTION
Diverse in tropics and subtropics, especially of Old World

IMPORTANT GENERA
Fulgoraecia and *Heteropsyche*

HABITAT
Tropical and temperate forests, woodlands, and shrublands

HOST ASSOCIATIONS
Obligate, external parasitoids of planthoppers, cicadas, and kin; so far as known, most species have narrow host ranges

CHARACTERISTICS
- Adult moths have wingspans from 0.2–2.2 in (5–55 mm); wings rounded
- Antennae short (less than half FW length); bipectinate to apex in both sexes
- Haustellum absent (adults nonfeeding)
- Pupation in a dense, white, chiton-shaped cocoon spun conspicuously on vegetation below the hopper's host plant

ZYGAENOIDEA: LACTURIDAE
TROPICAL BURNET MOTHS

This small family includes about 140 species parsed among 11 genera, with many more yet to be named.

The larvae and adults are boldly colored, which suggests that both stages are unpalatable. Adults can remain active in collecting jars charged with cyanide gas that would overwhelm most moths in seconds. It is thought that lacturids manufacture and discharge cyanide-laden defensive compounds, as do kindred burnet moths (Zygaenidae).

The larvae are particularly attractive animals, with bright pastel and black markings. Interestingly, their coloration derives from internal pigments, as the body wall of the thorax and abdomen is essentially transparent. The outside of the body is tacky to the touch and presumably serves to discourage would-be natural enemies. When molested, the caterpillars extrude bubble-like vesicles along the venter of the thorax and abdomen, adding to the stickiness of the caterpillar. Lacturids can be local pests, reaching densities that result in defoliation of their host.

ABOVE RIGHT | Tropical burnet moths like this *Lactura sapotearum* are virtually all boldly and brightly marked and likely unpalatable both as caterpillars and moths. The chemical nature of their defenses is in need of study.

INSET | *Enaemia pupula* caterpillar. Lacturid caterpillars have a thick, outer, tacky translucent layer that is a deterrent to many natural enemies. View a caterpillar under a lens to watch its dorsal heart in action.

DISTRIBUTION
Most diverse across Old World tropics and subtropics

IMPORTANT GENERA
Anticrates, *Gymnogramma*, and *Lactura*

HABITAT
Forests, woodlands, savannas, and shrublands

HOST ASSOCIATIONS
Woody plants; many associated with sapodilla family (Sapotaceae)

CHARACTERISTICS
• Brightly colored nocturnal moths with wingspans from 0.5–2.4 in (12–60 mm)

• Antennae filiform, about three-fifths of FW length

• Larvae gumdrop-like, brightly (evidently warningly) colored with tacky, transparent body wall; some extruding saclike vesicles along subventer when threatened

• Pupation in dense cocoon impregnated with whitish secretion

ZYGAENOIDEA: DALCERIDAE
JEWEL CATERPILLAR MOTHS

This is a small but extraordinary lineage of tropical moths, with a little over 80 described species and 11 genera endemic to the Neotropics. Their inclusion here anchors to their incredible caterpillars, which are among the most beautiful and strange insects on our planet and serve as a reminder of how much there is to discover about Earth's insect life.

The adults, nearly always conspicuously colored in white, yellow, or orange, are wonderful creatures: spindly and spiderlike in repose, with the "feathered" legs held forward.

The brightly colored larvae are largely transparent, with jellylike extrusions that detach if and when the caterpillar comes under attack (and prior to pupation). That they are shunned by birds, ants, and other would-be enemies seems to be a certainty, but little is known about the chemical or physical nature of their jellylike exterior. These are fascinating animals, about which we know far too little.

INSET | Dalcerid caterpillars have been likened to terrestrial sea slugs because of their beauty and mode of travel, as they, like other zygaenoids, glide rather than crawl.

LEFT | The adult resting posture of jewel caterpillar moths as shown here is diagnostic.

DISTRIBUTION
Neotropics; only a single species occurring north of Mexico

IMPORTANT GENERA
Acraga, *Dalcera*, and *Dalcerides*

HABITAT
Principally tropical forests

HOST ASSOCIATIONS
Poorly known; most thought to be generalists on woody plants

CHARACTERISTICS
- White, yellow, or orange moths commonly with wingspans from 0.8–1.6 in (20–40 mm); steeply declining wings held nearly vertically against body at rest
- Haustellum vestigial (adults nonfeeding)
- Larvae translucent with detachable jellylike accretions; highly reduced thoracic and abdominal prolegs

ZYGAENOIDEA: LIMACODIDAE
SLUG CATERPILLAR MOTHS

This is the largest of the 12 zygaenoid families, with more than 1,600 species and 300 genera worldwide. Limacodids are especially diverse in tropical forests, where their larvae astound with their variety and seemingly endless palette of colors.

The adults, which tend to be well represented at light collections, are often rendered in earth tones, but many are handsomely set with green or silvery scales; however, because of their smaller size, they often get overshadowed by the larger sizes, greater variety, and beauty of other tropical moths. Both sexes are rapid fliers, at least over short bursts, flying more like a bee than other moths. The haustellum is short or absent—the adults are essentially nonfeeding, with short life spans.

The caterpillars are often likened to sea slugs because of their beauty and legless-ness (they appear to glide across leaf surfaces) and because many sting. While most slug moth caterpillars are believed to be polyphagous, they commonly have local host plants that are favored.

Some slug moth caterpillars are protected by batteries of stinging setae—a trait that surely ties to their evolutionary success relative to other members of the superfamily. The stings of most species are on a par with those delivered by stinging nettles. Envenomization by some of the larger tropical species, however, especially in tender skin, can be quite painful, but none are considered a major health risk, unless an allergic reaction is triggered.

RIGHT | Limacodid larvae are amazing insects and make good quarry for caterpillar hunting and photography. More than half of the tropical species have unknown life histories. (All heads face to the left.)

LEFT | Many slug caterpillar moths have a distinctive habitus: the body is short and stout, with wings held almost vertically to either side of the body and the hairy legs held outward. Adults tend to be rendered in earthen colors or pastels, in contrast to their lavishly adorned caterpillars.

DISTRIBUTION
Cosmopolitan; diverse in tropics and subtropics; poorly represented across northern hemisphere

IMPORTANT GENERA
Acharia, Euclea, Latoia, Natada, Parasa, Perola, Scopelodes, and *Thosea*

HABITAT
Diverse in tropical and subtropical forests but generally widespread in woodland and shrublands dominated by woody plants

HOST ASSOCIATIONS
Very diverse, ranging from being dietarily specialized to exceedingly polyphagous; generally favoring smooth-leaved, woody plants; oaks support many species across North America; monocot hosts include palms, banana, and grasses

CHARACTERISTICS
• Small to medium-sized, thick-bodied moths with wingspans from 0.6–2.0 in (15–50 mm); most under 1.2 in (30 mm)

Saddleback caterpillar (*Acharia stimulea*)

Stinging Rose caterpillar (*Parasa indetermina*)

Hag Moth (*Phobetron pithecium*)

Monoleuca semifascia species complex

Doratifera quadriguttata

Euclea obliqua

- Robust thorax with short, triangular FWs, which are usually rounded apically
- Thick antennae with short, thick rami
- Adult posture characteristic with body held above substrate on densely scaled ("feathered") legs and steeply declining wings, appressed to sides of abdomen
- Larvae sluglike, nearly legless, with head pulled into thorax; many with stinging spines

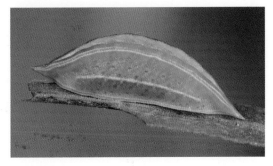

Perola clara

ZYGAENOIDEA: MEGALOPYGIDAE
FLANNEL MOTHS

LEFT | Megalopygidae caterpillars appear adorable, but their beauty ends and danger starts under "the plush" of silky, outer setae.

TOP RIGHT | *Megalopyge lanata* warns of its dangers with bold red, black, and white colors. While adults do not sting, the adults of some *Megalopyge* give off a foul odor if handled roughly.

MIDDLE RIGHT | No name was available for this flamboyant Peruvian megalopyid caterpillar at the time of publication—a reminder of how much is still unknown about the entomofauna of the tropics.

BOTTOM RIGHT | *Trosia* adults are decidedly warningly colored, but the defensive chemistry that sits behind this warning has yet to be elucidated.

Flannel moths, endemic to the New World, are surprisingly understudied relative to their medical importance, attractive adults, and flamboyant larvae. More than 240 species have been described but many more await description. Accruing molecular data suggest the family is rich in cryptic species.

The fetching adults are densely scaled, often in pastels, and frequently adorned with scale tufts over the thorax and along the abdomen. They are robust, strong fliers with a muscular thorax. Females have deciduous scales that are generously layered over the egg rafts as they are being deposited.

The larvae of some genera are among the "hairiest" of caterpillars, with densely packed, soft, hairlike setae covering all but the venter of the nearly legless caterpillars. Below the "toupée" of soft secondary setae, the caterpillars are armed with three rows of toxin-filled setae capable of delivering painful stings. Envenomization by megalopygids frequently results in acute discomfort for hours and soreness in upstream and downstream joints for days thereafter. The consequences of acute exposure to the large

DISTRIBUTION
New World, especially Neotropics

IMPORTANT GENERA
Megalopyge, Mesoscia, Norape, Podalia, and *Trosia*

HABITAT
Forests, woodlands, and scrublands

HOST ASSOCIATIONS
Larval diets tend to be generalized; almost exclusively tied to woody hosts; seemingly favoring plants with smooth leaves

CHARACTERISTICS
• Attractively scaled, nocturnal moths commonly with wingspans from 1.0–2.2 in (25–55 mm)

• Antennae short, bipectinate to apex in both sexes

• Densely and boldly scaled legs, held outward from body in repose, yielding spidery look

• Short haustellum (most thought to be essentially nonfeeding as adults)

Neotropical caterpillars, which can approach 2.8 in (7 cm) in length, have yet to be thoroughly documented, although there are rumors of stings leading to human death. Such instances, if true, may represent cases of an anaphylactic reaction rather than a typical response.

The conspicuously colored adults are also likely to be chemically protected. I have collected adults of several taxa in Costa Rica and Ecuador that emit a foul smell when threatened. Megalopygids tend to be dietary generalists but with local food-plant preferences.

- Larvae densely covered with soft setae that overlie rows of venom-containing setae
- Tough-walled cocoon with operculum (escape hatch)

ZYGAENOIDEA: HIMANTOPTERIDAE
LONG-TAILED BURNET MOTHS

Himantopterids are immediately distinguished by their highly modified hindwings that are drawn into long tails. The approximately 60 species are placed into 5 genera; some works include the Anomoeotidae as a subfamily of himantopterids. They are included here because of their phenotypic novelty and to draw attention to how little we know about their behavior, early stages, and host associations. The function of their tails remains a mystery.

It is unclear why himantopterids are generally rare in collections. Some early reports suspected an association with termites, but there are no recent confirmatory accounts of such. The handsome and phenotypically novel adults are thought to be primarily diurnal, but some are also seen at lights.

The larvae of some himantopterids are believed to be polyphagous. If so, they might be easily reared by confining gravid females and raising larval cohorts from eggs. Caterpillars of *Shorea* rest in large aggregations on the trunks of their hosts; larvae have also been observed to move en masse down the trunk of their host to pupate together underground.

ABOVE | One must travel to Africa or Southeast Asia to see these remarkable moths. Very little is known about their behavior, early stages, and host associations. While some Himantopteridae lack tails, most have tails 2–3 times their body length.

DISTRIBUTION
Tropical areas of Africa; a few species ranging into South Africa and Southeast Asia

IMPORTANT GENERA
Anomoeotes, Doratopteryx, Semioptila, and *Staphylinochrous*

HABITAT
Tropical and subtropical forests, woodlands, and scrublands

HOST ASSOCIATIONS
Poorly known; dipterocarps host some

CHARACTERISTICS
• Small to medium-sized moths with wingspans mostly 0.6–1.6 in (15–40 mm); wings often black with orange spots, some thinly scaled; HW extended to long tail

• Short, plumose antennae with thick rami in males; female antennae with short rami and scale tuft at apex

• Larvae small, stout, legless, sluglike, with abundant secondary setae from verrucae; triordinal crochets in two mesoseries on A3–A6

ZYGAENOIDEA: ZYGAENIDAE
BURNET OR SMOKY MOTHS

This extraordinary family includes many beautiful species with fascinating biologies. More than 1,000 species and 170 genera are distributed worldwide. They are exceedingly diverse in size and character, ranging from small moths with wingspans not reaching even half an inch to large tropical species with wingspans of 4 in (102 mm).

The wings are of such resplendent beauty that they eclipse all but the most beautiful butterflies.

The five subfamilies are so heterogeneous in nature that no single character unites them, and there is speculation that they may not represent a single family. For the purposes of this effort, the profile anchors to three core subfamilies: Chalcosiinae, Procridinae, and Zygaeninae, which include about 95 percent of the described species.

So handsome are the burnet moths (*Zygaena* and kin) of the Old World that they are often included in collections and books ostensibly about

ABOVE AND INSET | Chalcosiines include many of the most stunning moths to be found on Earth. All but a few of the 400 described species are endemic to Southeast Asia and Wallacea.

DISTRIBUTION
Cosmopolitan; diverse in tropics and subtropics, especially of Southeast Asia; rich fauna of burnet moths (Zygaeninae) in Palearctic; modest fauna in Nearctic

IMPORTANT GENERA
Cyclosia, Jordanita, Pyromorpha, and *Zygaena*

HABITAT
Diverse in nature, from drylands to wet forests; tropical and subtropical forests; chaparral, savanna, grasslands, and deserts

HOST ASSOCIATIONS
Broad array of hosts; many members of four o'clock, rose, and pea families

ABOVE | Chalcosine zygaenids, largely confined to Asia, Indo-Malaysia, and adjacent regions, are among the most beautiful of all moths. Their caterpillars can be equally striking in color. Shown here is the female of *Cyclosia papilionaris*; the males are stunning: plum colored with metallic green and blue detailing.

TOP LEFT | The orange and black adults of *Pyromorpha dimidiata* are members of a large and taxonomically diverse Müllerian mimicry complex that appears to be anchored to highly toxic net-winged beetles.

ABOVE | While *Jordanita chloros* lacks classical warning coloration, we are just beginning to understand how metallic scaling, UV reflections, and other types of signals are used by insects to signal to birds—the primary predators of day-flying moths.

CHARACTERISTICS
- Mostly day-flying moths with wingspans from 0.4–4.7 in (10–120 mm), with most well under 1.2 in (30 mm)
- Prominent antennae, often with short rami, usually between half and three-quarters of FW length; sometimes apically thickened (e.g., *Zygaena*)
- Larvae of three subfamilies with secondary setae, gathered onto swollen warts (verrucae); head withdrawn into prothorax

butterflies. Both larvae and adults are protected by self-manufactured, cyanide-generating precursors, and as such are often brightly or otherwise warningly colored. They serve as models in Batesian and Müllerian mimicry complexes across the planet. Adults of the three aforementioned subfamilies may release an acrid, cyanide-laden foam from the base of the eye and haustellum when alarmed. Likewise, the caterpillars are often boldly marked and conspicuous in habit.

Interestingly, many zygaenid host plants produce cyanide-releasing compounds when under attack as a way of discouraging their would-be herbivores. Zygaenid moths, however, are biochemically pre-adapted to handle such deterrents with aplomb. Larvae are frequently gregarious, at least through the early instars.

ZYGAENOIDEA
ZYGAENA BURNET MOTHS

Burnet moths of the genus *Zygaena* are an exceedingly handsome and popular group of diurnal moths. If measured by number, with more than 120 described species, the genus has been enormously successful evolutionarily (the average European moth genus has fewer than 10 species). Much of their diversity is endemic to arid and semiarid communities of the southern Palearctic, southwestern Asia, and northern Africa.

Relative to most day-flying moths, they are remarkably sluggish and lethargic, which, if nothing else, is a testament to their toxicity. Their cyanide-based defensive chemistry protects the early stages and adults from a sweep of natural enemies.

The larvae are host-plant specialists, primarily on members of the pea family (Fabaceae). Adults are conspicuous flower visitors and often co-occur with congeners. Many have quite restricted distributions.

LEFT | *Zygaena* has attained special status among insects in being handsomely colored, diurnal, and exceptionally diverse—indeed, *Zygaena* is the only moth genus in this book to get its own profile. Shown here is *Zygaena trifolii*.

ABOVE | Zygaenids typically are host-plant specialists on low-growing plants. The caterpillars are sluglike, with numerous, short secondary setae. Many are brightly colored, warning of their unpalatability.

DISTRIBUTION
Centered around Mediterranean region: southern Europe, Asia Minor, and northern Africa; a few species range into northern Europe and eastward into eastern Palearctic and Southeast Asia

HABITAT
Sunny, open communities, from deserts, grasslands, and steppes to meadows, chalk downs, and alpine meadows

HOST ASSOCIATIONS
Mostly tied to pea family; also buttercup, carrot, and mint families; some polyphagous

CHARACTERISTICS
• Black and/or red day-flying moths with wingspans from 1–2 in (25–50 mm); many with red, white, or creamy spots

• Antennae thickened distally (somewhat clubbed)

THYRIDOIDEA: THYRIDIDAE
WINDOW-WINGED MOTHS

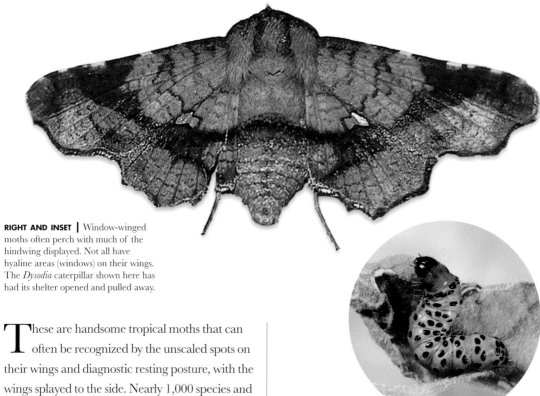

RIGHT AND INSET | Window-winged moths often perch with much of the hindwing displayed. Not all have hyaline areas (windows) on their wings. The *Dysodia* caterpillar shown here has had its shelter opened and pulled away.

These are handsome tropical moths that can often be recognized by the unscaled spots on their wings and diagnostic resting posture, with the wings splayed to the side. Nearly 1,000 species and about 95 genera are described. Thyridids are allied with the Hyblaeoidea and Pyraloidea but lack the well-developed abdominal tympanum of pyraloids. Most are nocturnal, but diurnality is common.

The larvae feed internally, often in tied leaves and flowers, and less commonly in fruits. Their shelters range from amorphous silken collections of plant tissue to intricately constructed rolls and spiral cones; a few form galls. Frass is often retained within the shelters and rolls.

Explosive regurgitation has been reported for a few species, wherein vomit is forcibly expelled well beyond the oral cavity. A few are occasional pests.

DISTRIBUTION
Richly represented in tropics; close to one third of described species occur in Africa

IMPORTANT GENERA
Arniocera, Banisia, Dysodia, Hypolamprus, Rhodoneura, and *Striglina*

HABITAT
Principally lowland tropical and semitropical areas; and temperate woodlands to deserts

HOST ASSOCIATIONS
Highly diverse, with host records from over 30 plant families

CHARACTERISTICS
• Wingspans from 0.5–2.8 in (12–72 mm); wings commonly with a few clear areas (windows) and reticulate patterning over upper and lower surfaces

• Perch with wings held to side and HWs displayed; head end elevated

• Haustellum short or absent; when present, without scales at base

• Larvae with only two L setae on T2 and T3

HYBLAEOIDEA: HYBLAEIDAE
TEAK MOTHS

This is a small tropical superfamily with a single family containing fewer than 20 species. The adults are stout-bodied, often with gray, tan, or red-brown forewings and concealed yellow to orange hindwings. Males have a large metatibial scent brush that is housed in process that extends down from the coxa. How the brush is used in courtship and mating has not been studied.

Hyblaeid caterpillars are commonly leaf-rollers and leaf-tiers in recently flushed leaves. *Hyblaea* larvae feed beneath leaf flaps that are folded over from a leaf edge and tied down with silk.

When alarmed, caterpillars of this family are able to explosively empty the gut contents in squirts on would-be attackers. *Hyblaea puera*, native to the Old World, was introduced into the Neotropics, where it has become a pest defoliator of catalpa, chaste tree, teak, and others.

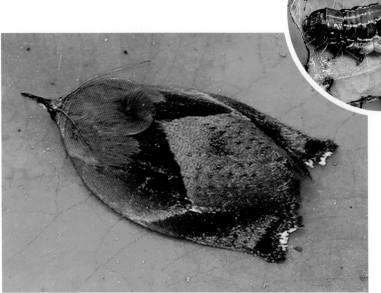

LEFT | Many hyblaeids share a common facies: medium-sized and stout, with a pronounced snout, small eyes, and squared-off forewings; the hindwings, covered in repose, often have bright orange markings. Shown here is *Hyblaea firmamentum*.

INSET | Last instar of Teak Defoliator (*Hyblaea puera*) shown inside its leaf shelter. Larger caterpillars and pupae of *H. puera* are eaten by some Indonesian cultures.

DISTRIBUTION
Mostly Old World; one endemic Neotropical genus

IMPORTANT GENERA
Hyblaea and *Torone*

HABITAT
Tropical and subtropical forests

HOST ASSOCIATIONS
Both specialists and polyphagous generalists; hosts include trumpet vine, mangrove, myrtle, and dipterocarp families

CHARACTERISTICS
• Stout moths with wingspans mostly from 1.0–1.6 in (25–40 mm); HW boldly colored; FW somewhat truncate

• Palpi projecting forward to form snout

• Coxal sheathlike extension enveloping massive tibial scent brush in males

• Lower surface of tarsi densely set with minute spines

• Larvae with bisetose L group on T1; crochets at least partially triordinal

CALLIDULOIDEA: CALLIDULIDAE
OLD WORLD BUTTERFLY-MOTHS

This is an intriguing group of mostly day-flying moths of the Old World tropics. The superfamily contains a single family with 50 or so species referred to 7 genera. Recent molecular data suggest the lineage is rather distantly related to true butterflies and instead more closely related to the window-winged moths (Thyatiridae).

The slender, attractively colored adults are butterfly-like in appearance and behavior: some perch with their wings held together over the thorax as in butterflies, although their filamentous, down-curving antennae, without a terminal knob, immediately distinguish them. And while they have a haustellum, they are rarely if ever seen at flowers taking nectar. Larvae feed within shelters on ferns, tying together or rolling pinnae.

LEFT | Callidulids, native to Madagascar and Southeast Asia (and adjacent areas), are a wonderful group of mostly day-flying moths. Many are aposematic or boldly patterned in white or black, which is suggestive that they are unpalatable.

INSET | Terminal (paragenital) courtship brush of *Pterodecta felderi*. While most moths only display their androconia during courtship—to protect the potency of the volatiles—moths with pheromones that include defensive properties will display their androconia when threatened.

DISTRIBUTION
Southeast Asia and Madagascar; a few reaching Palearctic Asia

IMPORTANT GENERA
Callidula and *Tetragonus*

HABITAT
Principally lowland tropical and semitropical communities

HOST ASSOCIATIONS
Poorly known; several are fern specialists

CHARACTERISTICS
• Medium-sized to moderately large moths; wings commonly held over body as in butterflies

• Antennae usually filiform; often thickened apically but not as distinctly clubbed as in butterflies

• Fourth tarsomere with pair of conspicuous apical spines

PYRALOIDEA: PYRALIDAE
PYRALID SNOUT MOTHS

More than 6,250 pyralid species have been described; these have been referred to over 1,000 genera. Enormous diversity remains undescribed from tropical areas. Five subfamilies are widely recognized—Chrysauginae, Epipaschiinae, Galleriinae, Phycitinae, and Pyralinae—the two largest of which are treated separately in the following profiles.

Pyralids are exceedingly diverse in appearance, making recognition of them difficult. Likewise, they are exceptionally varied in their biologies and include some of the most fascinating life histories exhibited by any moths. Given that most of their species diversity is in tropical systems, which are poorly explored, the family offers a galaxy of opportunity for study (and discovery).

ABOVE | Pyralids and crambids are successful, speciose, and diverse in habit, especially in the tropics where many have yet to be described. Most are relatively cryptic as adults, colored in earthen tones and pastels.

INSET | Many pyralid caterpillars, especially taxa that feed in stored products, bore into plants, or build shelters, tend to be rather mundane in both color and patterning. *Macalla thyrsisalis* is a notable exception, perhaps because its shelter is slight and the larva is visible within.

DISTRIBUTION
Cosmopolitan, from far north to far south

IMPORTANT GENERA
Refer to subfamily accounts

HABITAT
Especially tropics and semitropics but also temperate and desert communities

HOST ASSOCIATIONS
Exceptionally varied in part due to hyperdiverse nature of the family; feeding on all plant parts but also including fungivory, scavenging, and detritivory; some prey upon other insects

CHARACTERISTICS
• Small to large moths but most with wingspans under 0.8 in (20 mm)

• Abdominal tympanum recessed within a nearly closed cavity; conjunctiva and tympanum in same plane

• Larvae with SD1 seta on A8, usually with distinct sclerotized ring about its base; only two pre-spiracular L setae on T1

PYRALOIDEA: PYRALIDAE: PYRALINAE
PYRALINE SNOUT MOTHS

Taken as a whole, pyralines are confoundingly diverse in form and habit across the 1,320 or so described species. No adult or larval feature is known that uniquely unites the pyralines.

Some of the more brightly colored tropical taxa are thought to be mimics of unpalatable moths and beetles. The subfamily is best represented in the Old World, especially across Asia and Africa.

Life histories are not well known, perhaps because most are scavengers that feed on dead vegetation, seeds, bat guano, and other (mostly fallen) organic matter; several are stored-product pests on grains; others feed in galls and wasp nests. The nominate genus, *Pyralis*, includes important pests of birdseed, pelletized dog food, cereals, and other milled grain products.

LEFT | The resting posture of *Endotricha ignealis* shown here is typical of many pyraloids: the antennae are folded back and lie against the wings, and the anterior end of the body is modestly elevated.

DISTRIBUTION
Cosmopolitan, especially Asia and Africa

IMPORTANT GENERA
Aglossa, Dolichomia, Endotricha, Herculia, Pyralis, and *Vitessa*

HABITAT
Diverse, from tropics to many temperate communities; also grasslands and drylands

HOST ASSOCIATIONS
Varied; many specialist leaf feeders on woody plants, including both conifers and broad-leaved plants; most thought to be scavengers on nonliving substrates

CHARACTERISTICS
• Small to medium-sized moths but most with wingspans under 0.8 in (20 mm); roughly triangular FW; many with dark red, brown, or yellowish FW

• Labial palpi upcurved and held against face or directed forward to form snout

• Tympanum with secondary venulae usually absent

PYRALOIDEA: PYRALIDAE: PHYCITINAE
KNOT-HORNED MOTHS

This is a large, successful, cosmopolitan subfamily with more than 4,000 estimated species. Adults are usually gray with narrowly triangular forewings and have upcurved palpi. The anterior end of the body is elevated in repose. Adult males have a diverse and curious array of secondary sex (androconial) scales. While typically associated with abdominal scent pouches, androconia in the Phycitinae are also located on the wings, thorax, palpi, or antennae. When on the latter, they are recessed into a cavity, near the antennal base, whence the subfamily derives its common name.

The caterpillars are internal feeders in leaf shelters or rolls; many are borers in shoots, flowers,

ABOVE AND INSET | Most phycitines are gray and elongate-triangular when viewed from above, as in *Cactoblastis cactorum* shown here—a hero and a villain. A hero where it has been imported to destroy invasive stands of prickly pear cactus and a villain where it has colonized parts of North America, where there are over 200 species of native *Opuntia* cactus. Its caterpillar is shown in the inset.

buds, and cones; some form galls, while others are inquilines in galls of other species; a few are detritivores in ant nests; and still others are predators of scale insects. Included are some of the world's most notorious stored-product pests: for example, *Plodia interpunctella*, *Ephestia kuehniella*, *Cadra cautella*, and *Etiella zinckenella*.

DISTRIBUTION
Cosmopolitan

IMPORTANT GENERA
Cactoblastis, Cadra, Dioryctria, Ephestia, Ephestiodes, Homoeosoma, Peoria, Plodia, and *Phycitodes*

HABITAT
Diverse in deserts and grasslands; tropical, temperate, and boreal communities

HOST ASSOCIATIONS
Exceedingly diverse; including many specialists on broad-leaved plants, conifers, herbs, grasses, seeds, stored products, and still others

CHARACTERISTICS
• Mostly small moths with wingspans generally under 0.8 in (20 mm) but some to 1.3 in (33 mm); narrowly triangular FW

• Typically lack snout of other pyralids; instead with upcurved palpi

• Tympanum with circular thickening where nerves attach to membrane

• Larvae with SD1 seta on mesothorax with distinct sclerotized ring about base

PYRALOIDEA: CRAMBIDAE
CRAMBID SNOUT MOTHS

The more than 10,600 crambid species are placed across some 1,000 genera, with thousands from the tropics and regions of the southern hemisphere yet to be given a name. Fifteen subfamilies are widely recognized at present—four of these are treated in the following profiles.

The extraordinary richness and heterogeneity across the family will take decades to comprehensively assess—expect major changes in the higher classification as more taxa and genomic data are brought to bear on the family's phylogeny and classification.

Crambids display a seemingly infinite variety of structures, scent scales, and pheromone glands that come into play during courtship and mating; some even sing to one another as part of their courtship. Life histories are exceedingly diversified and include subterranean and aquatic lineages. While most are phytophagous, others are scavengers, inquilines in nests of ants and wasps, and/or predators.

The caterpillars are concealed feeders that form silk webs, galleries, and many types of leaf rolls and ties, typically with the larval feculae retained in the construction, seemingly for defense. The family includes dozens of important cereal, field crop, and orchard pests, especially in the Crambinae and Pyraustinae.

LEFT | Box Tree Moth (*Cydalima perspectalis*) is native to the Far East but established in Europe in 2009, where in less than two decades it became more common and widespread than in its home range.

DISTRIBUTION
Cosmopolitan

IMPORTANT GENERA
Refer to subfamily accounts

HABITAT
Virtually any community where insects occur

HOST ASSOCIATIONS
Highly diverse; mostly broad-leaved and coniferous plants but also graminoids, ferns, liverworts, mosses; feeding on all plant parts; some predaceous

CHARACTERISTICS
• Small to large moths with wingspans from 0.4–4.0 in (9–100 mm) but most from 0.6–1.0 in (15–25 mm)

• Abdominal tympanum with wide aperture; conjunctiva and tympanum in different planes, forming distinct angle where they meet

• Larvae with SD1 seta on A8 without distinct sclerotized ring about its base; only two pre-spiracular L setae on T1

PYRALOIDEA: CRAMBIDAE: CRAMBINAE
GRASS SNOUT MOTHS

While diverse in species, with likely over 2,100 described, it is their ecological abundance that most elevates the stature of crambines above that of many other moths. From the poles to the equator, in both dry and mesic, and from sea level to alpine grasslands, crambines can be among the most important insects, if measured by either biomass or ecological function.

The adults tend to be straw colored or other earth tones, with longitudinal markings that render them more cryptic when they are perched on grasses and sedges.

The caterpillars fashion elongate silken tunnels among the roots, stems, and leaves of their host grasses, sedges, and rushes. Many are internal borers in stems—included are a number of massive bamboo feeders. Crambines include important pests of turf grasses and a sweep of cereal crops: corn, maize, sugarcane, and rice. They include numbers of geographically restricted endemics with specialized habitat associations, and thus may make good bioindicators for conservation decisions regarding the management and restoration of grassy habitats.

LEFT | Grass snout moths such as *Crambus praefectellus* range from common suburban lawn denizens and occasional pests of sod and graminoid crops to highly specialized inhabitants of imperiled communities and ecosystems, and thus have much value as bioindicators of grass-dominated communities.

DISTRIBUTION
Cosmopolitan, from tropics to poles

IMPORTANT GENERA
Agriphila, Ancylolomia, Catoptria, Chilo, Crambus, Diatraea, Euchromius, Hednota, Pediasia, and *Prionapteryx*

HABITAT
Grasslands, marshes, and other environments dominated by graminoids

HOST ASSOCIATIONS
Most generalized grass feeders, including bamboo; many can be lab-reared on corn tassels; to what extent caterpillars use other plant matter is not well studied

CHARACTERISTICS
• Small to medium-sized moths with wingspans from 0.6–3.0 in (15–75 mm); FW elongate-triangular, often with outer edge notched or indented below apex; HW more than twice width of FW, with anal portion folding accordion-like at rest, with wings rolled around sides of body

• Snout especially long and in some genera hyperextended

PYRALOIDEA: CRAMBIDAE: ACENTROPINAE
AQUATIC SNOUT MOTHS

This is a successful subfamily of aquatic moths with over 800 described species found worldwide. Most can be recognized by their attractively patterned, meandering, mazelike wing markings.

At rest, acentropines typically expose some part of the hindwings. In the jumping spider mimics, the forewings bear pale markings that superficially resemble spider legs, and the exposed edge of the hindwings bears metallic spidery "eyespots."

The caterpillars of all but one genus are entirely aquatic, living within a case fashioned from silk and leaf fragments; those that inhabit fast-flowing waters live beneath a silken web spun under a submerged rock. A few are occasional pests in greenhouses and ornamental water features.

BELOW | Acentropinae are among the most varied, handsome, and curious microlepidopteran groups.

DISTRIBUTION
Cosmopolitan in tropical and temperate regions, including Oceania

IMPORTANT GENERA
Cataclysta, Elophila, Eoophyla, Nymphicula, Paracymoriza, Parapoynx, and *Petrophila*

HABITAT
Freshwater communities: ponds, lakes, and slow-moving streams and rivers favored but some inhabit fast-moving waters; *Nymphicula* are terrestrial case-makers in moist environments

HOST ASSOCIATIONS
Most generalized on submerged vegetation, including algae

CHARACTERISTICS
- Small to medium-sized moths with wingspans from 0.4–0.7 in (9–18 mm); wing patterns ornate; FW elongate

- Complex patterns of FW continue onto HW; caudal portion exposed at rest

- Most larvae immediately identifiable by their aquatic habit and gills (cuticular outgrowths) of the thorax and abdomen

PYRALOIDEA: CRAMBIDAE: PYRAUSTINAE AND SPILOMELINAE
WEBWORM SNOUT MOTHS

These two sister subfamilies were previously grouped together in Pyraustinae. About 1,300 pyraustines and 4,200 spilomelines are currently recognized.

As with other megadiverse insect lineages, the adults and larvae are marvelously varied in form and habit. While the great majority are leaf-tiers and leaf-webbers, others bore in stems, flowers, and fruits; some are inquilines and a few are predaceous. Their shelters and galleries tend to be messy: in design, the degree to which feculae are retained, and the use of silk within. The majority are dietary specialists as caterpillars, especially in the tropics and deserts, where plants tend to be well protected chemically.

Included are dozens of the world's most serious pasture and field crop pests. Conversely, many have been employed successfully for biological control of pernicious weeds (especially on islands).

LEFT | Many pyraustines like this *Pyrausta* are brightly colored and diurnal. Those scaled in pinks, reds, and yellows are especially pretty.

INSET | Larvae of the European Corn Borer (*Ostrinia nubilalis*) tunneling through a corn ear.

DISTRIBUTION
Cosmopolitan, including Oceania

IMPORTANT GENERA
Desmia, Diaphania, Herpetogramma, Omiodes, Ostrinia, Palpita, and *Pyrausta*

HABITAT
Exceedingly diverse, from tropics to temperate zones and subpolar regions; from forests and woodlands to grasslands, deserts, and other open habitats; from drylands to wetlands; some aquatic

HOST ASSOCIATIONS
Essentially all plant life; most on herbaceous and woody broad-leaved plants

CHARACTERISTICS
• Small to medium-sized moths with wingspans from 0.4–4.0 in (10–100 mm)

• CuP of FW often absent or represented by fold; veins 1A and 2A commonly fusing to form basal loop

• Male genitalia with gnathos usually vestigial or absent

MIMALLONOIDEA: MIMALLONIDAE
SACK-BEARER MOTHS

This is a small New World family with roughly 300 species and over 40 recognized genera, most of which are endemic to the Neotropics. Only five species occur north of Mexico; southern temperate areas have a richer fauna than the north, but most are tropical. Mimallonids are the sister group of the Macroheterocera.

The medium to large, stout-bodied adults are strong fliers. Some *Psychocampa* and at least one *Lacosoma* species are believed to be diurnal.

The early instars feed in a leaf fold, between two leaves, or under a reticulate silken net into which larval fecal pellets have been interwoven. In later instars, the caterpillars spin tough "sacks" out of leaf fragments and frass pellets, which are open at each end; both apertures can be simultaneously plugged with the highly modified abdominal terminus and the much sclerotized head. Their sacks can be secured to the host or freed to be made portable.

Even though they are concealed feeders, the caterpillars tend to be brightly colored, especially the portion of the body that is regularly extended out of the shelter to feed. Host-plant associations are incompletely known but most appear to be host-plant specialists on a spectrum of woody plants.

LEFT | Sack-bearers are robust moths with a proportionately thick thorax, small head, and prominent bipectinate antennae. If the perch allows, the wings are held downward, well below horizontal.

DISTRIBUTION
New World, with most endemic to tropics but extending into northern and southern temperate zones

IMPORTANT GENERA
Alheita, Bedosia, Cicinnus, Druentica, Lacosoma, and *Trogoptera*

HABITAT
Principally lowland tropical and semitropical communities; some ranging into temperate woodlands and higher elevations

HOST ASSOCIATIONS
Diversity of trees and shrubs; members of attorney tree, bushwillow, melastome, madder, myrtle, and sumac families commonly used; Nearctic species restricted to oak

Cicinnus melsheimeri

Lacosoma arizonicum

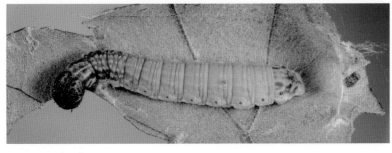

Lacosoma chiridota

LEFT | Sack-bearers are case-makers with the unique habit of being able to plug both ends of their case with a highly sclerotized head and posterior. The three terminal abdominal segments of *Cicinnus melsheimeri* (shown above) are fused to form an armored shield that prevents entry into the tail-end of the case. The three caterpillars shown here have all had their cases opened to allow for imaging. Otherwise, mimallonid caterpillars are quite cryptic in habit and rarely encountered.

CHARACTERISTICS
- Robust, thick-bodied moths with comparatively short wingspans from 0.8–3 in (20–70 mm); FW apex often falcate
- Nonfeeding adults with vestigial or absent haustellum
- Antennae usually bipectinate but rami narrowing abruptly or absent before terminus
- Hind tibiae lacking spurs
- Late instars with last two segments often forming shield used to plug rear entrance of case; head highly sclerotized and pitted; two L setae on T1; larval crochets arranged in transverse oval

MACROLEPIDOPTERA
MACROHETEROCERA OR MACRO MOTHS

Whether measured by number of species or biomass, ditrysian macrolepidopterans are the most ecologically important and evolutionarily successful Lepidoptera in forested ecosystems.

This section profiles five superfamilies. Currently, 26 families are recognized, but expect others to be added as some subfamilies are elevated to family status, southern hemisphere faunas are better sampled, and the phylogeny for macro moths gets further resolved. The phylogeny (opposite) shows evolutionary relationships among the 17 families profiled in this work. The terms macrolepidoptera, macro moths, macroleps, and Macroheterocera, as used in this book, are deemed equivalent. Papilionoidea (butterflies, including Hedylidae), Calliduloidea, and Mimallonidae—historically included in the macrolepidoptera—are excluded.

As implied by their name, macrolepidopterans tend to be larger moths with robust bodies. The CuP vein is absent in the hindwing. An ear, or structure sensitive to the ultrasonic frequencies of hunting bats, has evolved independently at least four times among Macroheterocera, which, if nothing else, attests to the central role that bats play in the night-to-night activities and diversification of moths. Drepanoidea and Geometroidea have independently acquired ears at the front of the abdomen; noctuoids have a metathoracic ear under the hindwing; and Sphingidae have an ultrasound detector at the apex of each labial palp. Macroheteroceran larvae share a unique arrangement of their crochets, with the hooklets running in a series roughly parallel to the body axis; in microlepidopterans, the crochets tend to be arranged in a circle or transverse oval (Mimallonoidea).

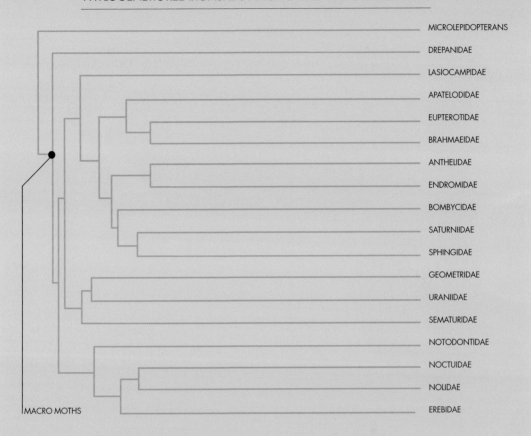

PHYLOGENETIC RELATIONSHIPS AMONG THE MACROLEPIDOPTERA

MICROLEPIDOPTERANS
DREPANIDAE
LASIOCAMPIDAE
APATELODIDAE
EUPTEROTIDAE
BRAHMAEIDAE
ANTHELIDAE
ENDROMIDAE
BOMBYCIDAE
SATURNIIDAE
SPHINGIDAE
GEOMETRIDAE
URANIIDAE
SEMATURIDAE
NOTODONTIDAE
NOCTUIDAE
NOLIDAE
EREBIDAE

MACRO MOTHS

Macrolep caterpillars are primarily external feeders, although a few lineages form shelters or nests; others are borers that tunnel into stem and root tissues; and a large number of Noctuidae are subterranean root feeders. Externally feeding caterpillars include some of the most interesting, colorful, and well-camouflaged insects. They account for much of the worldwide interest in moths and insect husbandry, provide many educational experiences that children have with insects, and are anchoring increasing numbers of ecology studies.

ABOVE | Inferred phylogenetic relations among 17 of the major families of macrolepidoptera, adapted from Kawahara *et al.* (2019).

DREPANOIDEA: DREPANIDAE
HOOKTIP MOTHS AND ALLIES

TOP | Many drepanids mimic fallen leaves with forewing patterns that are continued onto the hindwings, with the wings laid flat to either side of the body. Resting adults are seldom found in the daytime—it would be interesting to investigate where adults typically settle by day.

INSET | Several genera of drepanine caterpillars mimic rolled-up, dead leaves—the dragonesque *Oreta rosea* shown here is quite uncaterpillar-like.

The Drepanoidea include three currently recognized families: Drepanidae and two small lineages, the Doidae and Cimeliidae, both of which have fewer than 10 species. The 675 or so species of Drepanidae include about 125 genera that are parsed into 3 subfamilies: Cyclidiinae, Drepaninae, and Thyatirinae, although some genera are not well placed in any of the three.

The medium-sized to large adults are mostly nocturnal but are vexingly varied in habit and, as such, are difficult to characterize—one essentially needs to learn the three subfamilies to have a grasp of the family. A more thorough assessment of the phylogenetic relationships would add much clarity to their classification.

DISTRIBUTION
Cosmopolitan; especially well represented in Asia; poorly represented in Neotropics; absent from New Zealand

IMPORTANT GENERA
Cyclidia, Habrosyne, Metadrepana, Tethea, and *Thyatira*

HABITAT
Mostly temperate and tropical forests, woodlands, and shrublands; poorly represented in arid regions

HOST ASSOCIATIONS
Diversity of trees and shrubs, with host records from over 20 families; commonly feeding on members of birch, dogwood, honeysuckle, madder, myrtle, oak, rose, and willow families

ABOVE | *Drepana* have strongly hooked forewing apices. Their shelter-forming larvae use mandibular drumming and scraping as territorial signals to warn conspecifics of their location, as well as anal scraping using specialized setae.

Adults of all Drepanidae share an anatomically unique ear at the anterior of the abdomen, which allows them to detect echolocating bats.

The larvae, while also diverse in form and habit, can be immediately recognized by the possession of an extra L seta dorsocaudad of the spiracle. The caterpillars are host-plant specialists on woody plants, feeding externally or spinning a leaf shelter. Those of many thyatirines appear to be mimics of sawflies. *Epicampoptera* is sometimes a pest of coffee.

ABOVE | Thyatrines or false owlet moths were regarded to be a separate family until recent assessments revealed their phylogenic position within the Drepanidae. Much of their diversity is restricted to the northern hemisphere. Shown here is *Thyatira batis*.

CHARACTERISTICS
- Small to medium-sized moths with wingspans from 0.6–1.8 in (15–45 mm); FW often with apical lobe or falciform in Drepaninae; FW noctuid-like in shape and patterning in Thyatirinae
- Cavity at juncture of tergum and second abdominal sternum ("ear") enclosing two tympanal membranes
- Antennae filiform, serrate, or pectinate; usually with scales along dorsal side
- Haustellum present or absent; unscaled when present
- Diverse in form; always with additional L setae dorsocaudad of spiracle; anal plate drawn into short knob, and prolegs reduced without crochets in Drepaninae

LASIOCAMPOIDEA: LASIOCAMPIDAE
TENT CATERPILLARS OR LAPPET MOTHS

This is a moderately sized family with as many as 2,000 species and more than 150 genera parsed across 5 subfamilies. They are moderate-sized to large moths with broad wings and a stout body that is capable of powering the rapid flight of males and providing sufficient lift for the heavy, egg-laden females to fly. Lasiocampid adults tend to be densely scaled in earth tones or white and appear "hairy"; even the antennae are generously scaled. Such soft, hairlike setae absorb the ultrasounds emitted by hunting bats and help the moths to avoid detection. At rest, the leading edge of the hindwing often extends beyond the forewing as shown in the photograph below.

LEFT | *Chrysopsyche lutulenta* is a striking moth, seemingly both cryptic and aposematic. As with other lasiocampids, having an excessively hairy body makes it unsuitable prey for many natural enemies.

DISTRIBUTION
Cosmopolitan, with greatest diversity in tropics but reaching to boreal zones of both hemispheres

IMPORTANT GENERA
Artace, Euglyphis, Gloveria, Malacosoma, Tolype, and *Trabala*

HABITAT
Principally lowland tropical and semitropical forests and woodlands; sclerophyll communities and other shrublands; some adapted to drylands

HOST ASSOCIATIONS
Mostly generalists on woody plants, at least through later instars; some specialists in first and early instars; angiosperms primarily but many including conifers as their preferred hosts; chondrostegines generally tied to low-growing herbaceous plants and even consume grasses

CHARACTERISTICS
• Broad-winged with stout thorax; wingspans from 0.8–7.0 in (20–175 mm);

Lacking a haustellum and other functional mouthparts, the adult stage is wholly given to matters of reproduction and dispersal. Males are occasionally diurnal, with a restricted period of activity coinciding with the time of day when virgin females are calling—that is, releasing their sex pheromone. (Good luck trying to catch a seeking male, as the flights of the day-active species can be blisteringly fast.)

Tent-building lasiocampids are among the most social Lepidoptera, frequently living with their siblings through the entirety of their larval stage. Perhaps surprisingly, the majority of tent caterpillars are solitary and include some of the planet's most cryptic caterpillars: being bark-resters with shadow elimination hairs that render them nearly impossible to find by eye. A goodly number are positioned between these extremes; they are more or less cryptic at a distance but brightly colored in hand, usually marked with oranges or yellows that warn they are best left alone. Others have bright colors that are only displayed when alarmed: such as black, mouthlike markings hidden in the caterpillar's intersegmental folds that are revealed only upon disturbance, and/or aposematic venters that are waved to would-be attackers, when the caterpillar raises its anterior end to display its bright orange and black belly, while methodically panning its body from side to side.

Their nests or tents vary greatly in architectural design. In the Forest Tent Caterpillar (*Malacosoma disstria*), the caterpillars merely spin a sheeting of silk over a tree trunk that serves as the communal

ABOVE | *Gloveria* are large moths, with caterpillars that reach lengths of 4–5 in (100–125 mm). Male lasiocampids tend to have relatively short, plumose antennae specialized for the detection of females—no offense intended, but their eyes, brain, and head are rather puny.

wings often rendered in earth tones or white; HW small with enlarged humeral region, lacking frenulum
- Nonfeeding adults with vestigial haustellum
- Caterpillars with abundant, soft, secondary setae even across head capsule; bark-resters often with lateral lappets and brightly colored venters; crochets biordinal (of two alternating sizes); wartlike protuberance on venter between anal prolegs

roost for the larvae. The filmy silken tents of many tent caterpillars are remarkable in their ability to function as biological greenhouses that heat up much above ambient temperatures by day, and as such serve (among other things) to greatly speed up larval development in the early spring when the early instars are active.

The caterpillars' pelage of fine setae serves to discourage many bird and mammal predators, and because of such, may link to the flashy coloration of the tent-makers. While lasiocampid setae are nonstinging, some bear short, sharp, deciduous setae that can cause modest discomfort, itchy rashes and, in sensitive individuals, lead to unpleasant dermatological reactions. *Gloveria* and others have thousands of such setae along the dorsum and should not be handled unless done so guardedly—the setae will not penetrate the tougher skin of one's hand if the animals are held and corralled gently but can be problematic when lodged in more sensitive skin or if one is exposed to the setae in great number.

LEFT | Head of Forest Tent Caterpillar (*Malacosoma disstria*). Lasiocampids have setae over their whole body, including an abundance of short setae on their head capsule. While the long, silky hairs of tent caterpillars are nonstinging and non-urticating, tent caterpillars are avoided by many predators.

ABOVE | The Forest Tent Caterpillar is both a chronic and outbreaking defoliator of many deciduous forest types: black gum and aspen woodlands can be especially impacted. The nonfeeding adults are essentially volant reproductive vessels.

AMERICAN TENT CATERPILLAR

LEFT | Communal nest of *Malacosoma americanum* with nestmates tightly grouped outside their refugium, presumably thermoregulating. Last instars get increasingly solitary.

The American Tent Caterpillar (*Malacosoma americanum*) is among the most well-studied caterpillars in the world. Females lay their black, foamy egg clusters only on Black Cherry (*Prunus serotina*); but later instars become increasingly polyphagous, and will feed on many woody plants.

Interestingly, embryos of the American Tent Caterpillar develop almost to the point of hatching but forego eclosion for 9 or 10 months, holding in a state of suspended animation while awaiting the arrival of spring. As soon as the new cherry leaves begin to flush in the early spring, larvae are prepared to emerge and begin to form their communal nest, which will contain many dozens of caterpillars. When temperatures are cool, the caterpillars leave the tent to forage but return to the nest to digest their meals.

The filmy, silken tents are remarkable in their ability to function as biological greenhouses that heat up much above ambient temperatures by day and as such serve, among other things, to greatly speed up larval development at temperatures too cold for most caterpillars. (Old cherry leaves, once loaded with an array of secondary plant compounds, are lethal to tent caterpillars.)

Interestingly, during each foraging bout, a small fraction of the caterpillars serve as scouts that lead their siblings to and from foraging sites by laying down a trail of silk that is impregnated with a pheromone.

Lasiocampid nests can be a maze of tunnels, cast fecal pellets, and shed skins that block and confuse the entry of natural enemies, including birds and many mammals. But not all enemies: perhaps because of their abundance and conspicuousness, these nest-builders are heavily parasitized by tachinid flies and ichneumonid wasps.

When larvae of *Malacosoma americanum* are inadvertently consumed by grazing horses, the hairs may embed along the alimentary tract and lead to bacterial infections that can trigger abortions—when multimillion-dollar racehorses fall victim, the matter can make national news.

As is too often the case in the Anthropocene, this species has become decidedly less common during the past few decades over much of its range.

BOMBYCOIDEA: APATELODIDAE
APATELODID MOTHS

The higher classification of Bombycoidea is in great flux, with much disagreement as to how many and which families and subfamilies should be recognized. Ten bombycoid families are currently recognized, eight of which are featured in this book.

Included among Bombycoidea are most of the large to very large moths and many of the most familiar and loved insects and, of special commodity importance, the Domesticated Silk Moth. Apatelodidae, Brahmaeidae, and Eupterotidae form a clade that is sister to the other seven bombycoid families. The families are introduced here roughly in the order in which they diverged from other bombycoids, beginning with Apatelodidae.

The 12-plus genera and 180 species of described apatelodids are medium-sized moths that are best known for their distinctive leaf-mimicking postures and handsome larvae. The family is anchored to the Neotropics, with only a handful of species extending into temperate regions.

The nocturnal, nonfeeding adults have a diagnostic posture: the wings are held out to the side, but above the substrate, often asymmetrically, with the hindwings tucked out of view; many raise the abdomen, perhaps to resemble a twig.

ABOVE RIGHT | Apatelodid caterpillars are densely hairy, which evidently renders them unpalatable to most birds, even when the setae are non-urticating. *Olceclostera seraphica* shown here is a cryptic bark-rester.

RIGHT | Many tropical apatelodids, such as this *Prothysana*, are flamboyantly colored. It remains to be seen if they are mostly palatable Batesian mimics of toxic tiger moth caterpillars or unpalatable in their own right.

DISTRIBUTION
Neotropics, with small numbers extending into southern and northern temperate areas—six occur north of Mexico and only two into southern Canada

IMPORTANT GENERA
Apatelodes, *Olceclostera*, and *Prothysana*

HABITAT
Forests, especially through lowland tropical and semitropical communities; woodlands and shrublands; some extending into deserts and other drylands

HOST ASSOCIATIONS
Both dietary generalists and specialists; primarily using woody plants but some including herbaceous species

CHARACTERISTICS
- Medium-sized moths with wingspans often from 0.8–2.0 in (20–50 mm); FW triangular, often with outer margin concave below apex
- First two pairs of legs densely scaled and held outward and apart when perched

The caterpillars are flamboyant creatures vested in bright setae often, additionally, with dramatic tufts, lashes, and feather-like plumes, all of which are more or less deciduous and presumably protect the caterpillars. None are serious pests.

BELOW | Apatelodids can be recognized by their triangular forewings that are often scalloped below the apex. In repose, the first pair of legs is held forward and the second pair splayed to the side.

BOTTOM | To decide whether a Neotropical caterpillar is an arctiine tiger moth or an apatelodid, check the prolegs—they are often red in apatelodids, and the setae lack the minute barbs of tigers.

- Larvae with dense, silky, secondary setae; individual setae lacking barbs, as in tiger moths (Erebidae: Arctiinae); often with red prolegs; crochets biordinal in ellipse that runs parallel to body axis

BOMBYCOIDEA: EUPTEROTIDAE
MONKEY MOTHS

This is a modestly diverse family with fewer than 350 species and 35 genera. The family is both anatomically and biologically heterogeneous—a matter that is reflected in the instability of its classification. Various treatments recognize between three and six subfamilies, with several genera that don't fit well into any of these.

The adults range from small bombycoids to very large moths, with females averaging larger in size. The large-bodied, nonfeeding, densely hairy adults are typically rendered in pastels and earth tones with unassuming markings. As might be guessed from the coloration, the adults are mostly nocturnal. At rest, the wings are held to the side with the forewing covering about half the hindwing.

The caterpillars are densely setose and have earned members of the family a less familiar common name: the giant lappet moths. Their pelage of silky setae borders on extravagant and appears intended as an advertisement of their unpalatability—most are in no way cryptic and many include bright yellow, orange, red, and black setal lashes and tufts. Members of the Panacelinae have urticating (but not stinging) setae. A few are minor pests; *Nisaga simplex* is a rice pest in India.

TOP | Monkey moths are large bombycoids given to earthen colors and pastels (usually yellows). Shown here is *Eupterote mollifera*.

INSET | As exemplified here in *Sphingognatha asclepiades*, Eupterotid caterpillars are densely vested in strikingly long, silky setae.

DISTRIBUTION
Diverse across Africa and throughout Asia

IMPORTANT GENERA
Cotana, Eupterote, Gonojana, Jana, Piala, and *Stenoglene*

HABITAT
Tropical and subtropical forests, woodlands, and shrublands

HOST ASSOCIATIONS
Both dietary specialists and generalists; nearly always on woody plants

CHARACTERISTICS
• Small to very large moths with wingspans mostly from 1.0–5.5 in (24–140 mm); wings broadly triangular; FW apex sometimes outwardly produced

• Distal tarsomere of mesothoracic leg with medial row of spines

• Larvae with dense, silky setae; even the head bears abundant (shorter) setae; some taxa with barbed, spinose, and urticating setae as well; crochets biordinal

BOMBYCOIDEA: BRAHMAEIDAE
BRAHMIN OR OWL MOTHS

LEFT | Brahmin moths are Old World bombycoids: large, robust moths, with elaborate bipectinate antennae, and highly variable patterning.

INSET | Some genera of brahmaeid caterpillars are endowed with prominent scoli and horns; others have long primary setae amongst numerous, short secondary setae.

This is a small family with approximately 50 species and 6 genera. The family is strongly supported as the sister group to the Eupterotidae opposite. Much of their species diversity is endemic to Africa, where 38 species have been described, although some of these likely represent synonyms.

The adults tend to be large moths, colored in earth tones; most are marked with a series of parallel black lines on both the forewing and hindwing. Nearly all are nocturnal. The wings are held to the side, with about half the hindwing exposed. As in other bombycoids, the females are bigger than the males, with greatly enlarged abdomens.

The caterpillars may have extravagant horns (scoli), recurved or coiled at their apices, superficially resembling the larvae of some ceratocampine silk moths (Saturniidae). In contrast to the adults, caterpillars tend to be boldly marked, often with yellow or orange. Those of the nominate genus will roll their head under an enlarged thorax to yield a snakelike effect.

DISTRIBUTION
Africa primarily; a few species found across southern Palearctic and in Asia

IMPORTANT GENERA
Brahmaea, Dactyloceras, Lemonia, and *Sabalia*

HABITAT
Tropical and subtropical forests, mesic and xeric woodlands, and savannas

HOST ASSOCIATIONS
Incompletely known; most records from dogbane (Apocynaceae) and olive (Oleaceae) families

CHARACTERISTICS
• Medium-sized to large moths; wings broadly triangular

• Haustellum present or vestigial

• Abdominal segments 4–6 in adult with spiracles offset upward

• Larvae with T2 and T3 enlarged, often bearing paired dorsal horns (scoli); single medial scolus over A8

BOMBYCOIDEA: ANTHELIDAE
ANTHELID MOTHS

This is a small family, with 94 species distributed across 9 genera endemic to Australia and New Guinea.

Adults are large, thick-bodied, and densely covered with setae. Females are burdened with enormous abdomens; those of *Pterolocera* have foregone flight altogether in favor of greater reproductive output—the head, thorax, legs, and vestigial wings are comically small relative to the female's egg-packed abdomen. The adults tend to be tan, brown, and other earthen tones, with simplified pattern elements that cut through both sets of wings, serving to blend with the bark of trees and forest floor litter.

The larvae are densely covered with secondary setae: many bear long, hairlike setae that are loosely grouped on each segment as well as an undercoat of shorter setae that may be barbed, plumose, spinelike, or spatulate. While the setae do not sting, they can be urticating if the caterpillars are handled. The head is often boldly marked with a broad black band to either side of the frontal triangle.

LEFT | Anthelids are so variable that they are difficult to characterize as adults and larvae. They are large, often earthen colored. Shown here is *Anthela denticulata*.

INSET | Caterpillars are large, with abundant secondary setae or spines. Shown here is *Chelepteryx collesi*.

DISTRIBUTION
East Australia primarily, with modest representation into New Guinea

IMPORTANT GENERA
Anthela and *Chelepteryx*

HABITAT
Forests and woodlands

HOST ASSOCIATIONS
Most associated with myrtles, especially eucalyptus; generally polyphagous; some feed on grasses and exceptionally damage cereal crops

CHARACTERISTICS
• Medium-sized to large moths with wingspans mostly from 1.0–6.5 in (24–165 mm); outer margin of wings often scalloped; FW apex often drawn to acute angle, sometimes falcate

• Haustellum absent except in three species of Munychyiinae

• Males possessing wart rearward of spiracle on A1

• Larvae with dense secondary setae: long, and silky to shorter and urticating

BOMBYCOIDEA: ENDROMIDAE
GLORY MOTHS

This family is in a state of renaissance, with its contents changing as more molecular data are brought to bear on the phylogeny of the Bombycoidea. What until recently was thought to be a family with just one or two species has grown to include some 70 species. Recent molecular data indicate that endromids form the sister group to two small bombycoid families not treated in this work: Phiditiidae (23 Neotropical species) and Carthaeidae (one Australian species), which collectively are sister to Anthelidae (opposite).

Endromidae are one of the most morphologically heterogeneous taxa in the superfamily—no trait is known to unite its 16 genera.

Endromid larvae are fantastically diverse in form and habit. Larvae of several genera have fanciful lateral flanges, and many have a fleshy, medial horn on A8—words can scarcely do justice to these wonderful caterpillars.

TOP | The Kentish Glory (*Endromis versicolora*) is one of the most iconic moths across Europe and Asia. Its larva feeds on birch. Like many large moths, it is believed to be declining.

RIGHT | *Andraca theae*, endemic to Southeast China and Taiwan, is gregarious through the early instars. It is sometimes a serious pest of tea.

DISTRIBUTION
Tropics and semitropics of Southeast Asia; several species extending into Palearctic

IMPORTANT GENERA
Andraca, *Mustilia*, *Mustilizans*, *Oberthueria*, *Prismosticta*, and *Smerkata*

HABITAT
Tropical and temperate forests and woodlands

HOST ASSOCIATIONS
Mostly polyphagous on woody plants

CHARACTERISTICS
• Medium-sized to large, thick-bodied moths; females with large abdomens—many resembling bombycoids in form and habit; others fold wings over abdomen in tentiform fashion

• Haustellum vestigial

• Larvae exceedingly diverse in form and habit; secondary setae short and usually inconspicuous; several genera with lateral flutes and fleshy horn on A8

BOMBYCOIDEA: BOMBYCIDAE
SILK WORMS OR MOTHS

This is a small bombycoid family with about 200 species referred to 27 genera. Much of their species diversity anchors to Indo-Australia. Included is the world-renowned Domesticated Silk Moth (*Bombyx mori*), featured on page 180. Bombycidae are sister to Saturniidae and Sphingidae.

The adult resting posture is often diagnostic: the forewings are slightly raised above horizontal and held to the side; the hindwings are tucked under the forewings with the hind margin rolled upward; and the abdominal terminus is upcurved.

The caterpillars, being diverse in form, coloration, and habit, are difficult to characterize: they lack the long secondary setae and scoli of other Bombycoidea and instead have numerous, soft, secondary setae that are easily overlooked. Many have bold markings that suggest at least some are unpalatable, a possibility that is reinforced by the gregarious nature of the caterpillars of many species (palatable insects are seldom gregarious); further, their favored host plants, mulberry and cecropia, are known to be chemically protected and shunned by caterpillars with generalized diets, and unlike most bombycoid families, Bombycidae tend to be host-plant specialists.

Studies of the mating behaviors of *Bombyx mori* led to the chemical identification of the first animal

RIGHT | The sex scent released by a virgin Domesticated Silk Moth (*Bombyx mori*) was the first animal pheromone to be chemically characterized—and then dubbed bombykol. Many thousands of pheromones have since been identified from other moths, insects more broadly, and a sweep of vertebrates. Pheromones are particularly important for communication in animals that cannot rely on vision.

DISTRIBUTION
Pantropical, with richest representation in Asia and Indo-Australia extending west into Palearctic Asia, tropical Africa, and Madagascar; Epiini endemic to Neotropics

IMPORTANT GENERA
Bivincula, Bombyx, Elachyophtalma, Gunda, Ocinara, Racinoa, Anticla, Colla, Epia, and *Quentalia*

HABITAT
Tropical and subtropical forests

HOST ASSOCIATIONS
Bombycini closely tied to cecropia, figs, and mulberries

CHARACTERISTICS
• Medium-sized, thick-bodied moths with wingspans mostly from 0.8–3.7 in (20–95 mm); wing scales with acute toothed apices

• Females with large abdomens; terminal abdominal segment often upcurved at rest

pheromone, which was subsequently named bombykol. Later studies have found that bombykol is the main constituent of a multiple-constituent sex pheromone released by virgin females and that such pheromone blends that add signal complexity are the norm among moths, especially where multiple congeners live together, and the risks of a mating mistake are heightened.

ABOVE | Many bombycids, such as this *Epia muscosa*, have a peculiar resting posture: the forewing is held outward with the hindwing hidden beneath, the head is directed downward, while the trailing portions of the hindwings are elevated.

INSET | Caterpillars of at least four families of moths have independently converged on the stratagem of mimicking a bird dropping, especially through earlier (smaller) instars. Shown here is a middle instar of *Bombyx mandarina*.

LEFT | *Anticla antica* is highly variable in color and pattern. Such is common in many moths but virtually absent in butterflies, perhaps because visual communication rules in diurnal Lepidoptera, while smell is the primary communication channel for moths.

- Antennae relatively short; haustellum vestigial; maxillary galeae bladderlike in males
- Legs densely scaled and hairy; conspicuously displayed when moth is perched
- Larvae with short secondary setae; often marked with black bands or enlarged black pinacula; verrucae absent; commonly with middorsal horn or hardened plate on A8

MOTHS AND SILK PRODUCTION

LEFT | The Domesticated Silk Moth (*Bombyx mori*) has indeed lost its wildness: it can be raised in trays without enclosure, will accept leaves and conditions that would be shunned by most caterpillars, and it all but lacks defensive behaviors.

Seven species of *Bombyx* are endemic to East Asia and India. While both the Domesticated Silk Moth (*B. mori*) and the Wild Silk Moth (*B. mandarina*) are used in commercial silk production, the industry is anchored by *B. mori* and its many varieties. The Domesticated Silk Moth sustains the world's multibillion-dollar-per-year silk industry. The global silk market was valued at US$17.2 billion in 2022 and is projected to reach a value twice that by the end of the next decade.

While subspecific entities are often named across many groups of Lepidoptera, usually based on different wing markings, geographic ranges, host-plant usage, or, more recently, genetic differences, the named varieties of the silk moth are based on cocoon colors—a matter that translates to different textile colors and properties.

Bombyx mori has been in culture for some 5,000 years. The species was domesticated in China from *B. mandarina*. The moth has been in culture for so long that it is no longer capable of flight, and both the caterpillar and adult have become simplified in their patterning and commitment to pigment production.

Commercial silk comes from the moth's cocoon, which is constructed from a single fiber released by the prepupal caterpillar as it spins its refugium—the fiber is believed to reach c.2,000–3,000 ft (600–900 m) in length. By boiling the cocoon (with the pupa inside), it is possible to unwrap the cocoon without breaking the fiber. Silk threads are made by unraveling the cocoons and wrapping the silk fiber from each cocoon into a single bundle (thread) that can then be used to manufacture sheets, lingerie, kimonos, saris, sleeping bags, parachutes, bike tires, and lampshades and other furnishings. Caterpillar silk is seeing increased use in surgeries—for example, to provide a sterile scaffolding for wound healing after burns and plastic surgery.

BOMBYCOIDEA: SATURNIIDAE
EMPEROR AND GIANT SILK MOTHS

This is the most diverse of the 10 Bombycoidea families, with more than 3,450 species and 180 genera grouped into 7 subfamilies, 4 of which are introduced separately in the profiles on the following pages. Their species diversity peaks in the foothills of the Andes.

Collectively, they form the sister group to Sphingidae. Giant silk moths include many of the most familiar, beautiful, and beloved insects, as well as many of the largest, and as such are allotted in this work more pages than any other family of similar species richness.

As might be guessed from a taxon with seven subfamilies, saturniids are quite diverse in character and habit. The adults are mostly nocturnal, but diurnality has evolved independently across many lineages. While nocturnal taxa tend to be cast in earthen tones, family members also include fetching, brightly colored moths and are a

BELOW LEFT | *Citheronia regalis* is North America's most massive insect, with caterpillars sometimes exceeding 5 in (130 mm) in length.

BELOW RIGHT | There are 28 species of western North American *Hemileuca*. Most are diurnal and fast-flying. Baiting males with a virgin female is an effective way to see them.

DISTRIBUTION
Cosmopolitan, with greatest diversity in Neotropics

IMPORTANT GENERA
Refer to subfamily accounts

HABITAT
Tropical, semitropical, and temperate forests; sclerophyll woodlands and shrublands; deserts and other dryland communities

HOST ASSOCIATIONS
Larval diets run from broadly polyphagous to highly specialized; most polyphagous on woody plants, although generalists commonly have locally specialized preferences; vast majority on angiosperms, including grasses, but also gymnosperms

perennial favorite not only with collectors but also with breeders, educators, children, and insect aficionados. Legions of professional entomologists and other biologists have childhoods that trace to raising local species of these magnificent insects.

Their marvelously diverse larvae generally have scoli; those of Hemileucinae are hollow and loaded with venoms that can deliver memorable but rarely dangerous stings. Saturniid caterpillars change markedly in form and color across instars. Larger species are eaten by indigenous peoples and in some countries represent a significant food resource. A handful are economically important pests of timber.

ABOVE | Southeast Asia's magnificent Atlas Moth (*Attacus atlas*) is one of the largest moths in wingspan and mass. Wingspans frequently reach up to 11 in (280 mm).

INSET | Larvae of the Atlas Moth are lovely animals. Caterpillars may secrete considerable wax, yielding a curiously handsome blue-green flocculence to their dorsum.

CHARACTERISTICS
- Medium-sized to very large, thick-bodied moths with wingspans from 0.8–11 in (20–280 mm); in flight, wings held together by expanded humeral area of HW; frenulum absent; some with spectacularly tailed HW
- Abdomens ranging from proportionately small relative to wingspans (e.g., many Saturniinae) to large (e.g., Ceratocampinae)
- Male antennae large, bipectinate to quadripectinate
- Haustellum vestigial and adults nonfeeding
- Larvae exceedingly diverse in form and habit; secondary setae may be short and inconspicuous, but usually evident above prolegs; most bear scoli for one or more instars; anal prolegs enlarged and serving as claspers

BOMBYCOIDEA: SATURNIIDAE: SATURNIINAE
EMPEROR OR WILD SILK MOTHS

This is the second-largest saturniid subfamily, with over 630 recognized species and 50 genera. Like many other bombycoids, species diversity is greatest in the tropics and semitropics.

Saturniines are handsome creatures, marvelously diverse in appearance both as adults and as larvae. Included are some of the largest insects, such as *Attacus atlas* (opposite), with wingspans that commonly exceed 4 in (96 mm). They are also record holders in having the largest (measured by surface area) and most chemically sensitive antennae among animals.

Saturniines are easily the most popular, oft collected, and photographed of all moths. Their caterpillars, almost infinite in variety, are especially popular among insect breeders, children, students, and educators. Many are imperiled and receive conservation protection. Larger species are eaten by indigenous cultures, with the most well-known being the Mopane Worm (*Gonimbrasia belina*) (see page 42).

BELOW | The subtle browns, tans, and pastels common to moths yield some of nature's most stunning winged tapestries, as shown by this Giant Peacock (*Saturnia pyri*).

DISTRIBUTION
Cosmopolitan, with greatest diversity in South America

IMPORTANT GENERA
Actias, Antheraea, Imbrasia, and *Rothschildia*

HABITAT
Forests, woodlands, shrublands, and other communities with woody vegetation

HOST ASSOCIATIONS
Extremely diverse; most polyphagous on woody broad-leaved plants (angiosperms), although often with local host preferences

CHARACTERISTICS
• Medium-sized to very large moths; FW often with discal eyespot, lunate spot, or hyaline window

• Male antennae large, quadripectinate

• Mesoscutum without anterior middorsal projection

• Caterpillars with sparse secondary setae; these most evident above prolegs

BOMBYCOIDEA: SATURNIIDAE: CERATOCAMPINAE
ROYAL MOTHS

This is a large New World subfamily with more than 170 species distributed among 31 genera. It is a largely Neotropical radiation, with fewer than 30 species extending north of Mexico.

The forewings are often attractively colored; the hindwings are commonly rendered in yellows, oranges, and reds; eyespots are usually absent. There is often a single or paired white or black spot at the end of the discal cell of the forewing.

The males of many genera are strong fliers, approaching sphingids in power of flight. Most are nocturnal; males of *Anisota* are mostly day-flying.

The splendid, dragon-like caterpillars bear prominent scoli, with those of the thorax giving the larvae impressive horns; some reach massive sizes. The caterpillars tend to be specialists, at least locally, although many, and especially larger taxa, tend to be more generalized in diet. Gregarious taxa, such as *Anisota*, can be local defoliators of oak.

LEFT | *Psilopygida* are lovely saturniids endemic to South America, typically rendered in pinks, yellows, or gray scaling. Note how the hindwing extends beyond the forewing at rest.

INSET | *Syssphinx* are among the most beautiful and sought-after New World caterpillars (here, *Syssphinx blanchardi*). The adaptive significance of their silvery scoli has yet to be explained. Similar silver markings occur in a few Notodontidae.

DISTRIBUTION
New World

IMPORTANT GENERA
Adeloneivaia, Anisota, Citheronia, Eacles, and *Syssphinx*

HABITAT
Tropical forests, woodlands, tropical scrub and thorn communities, and savannas

HOST ASSOCIATIONS
Very diverse, with both dietary specialists and generalists; temperate genera often on members of oak and pea families

CHARACTERISTICS
• Large, robust moths with wingspans from 0.8–6.5 in (20–165 mm); FW apex often drawn into a point

• Head proportionately small relative to robust thorax and abdomen

• Antennae relatively small with distal portion lacking rami

• Caterpillars bear prominent scoli, often with those of thorax developed into elongate horns

BOMBYCOIDEA: SATURNIIDAE: ARSENURINAE
ARSENURINE MOTHS

This is a small subfamily with about 90 recognized species across 10 genera. They tend to be large moths, rendered in mouse gray, tan, and other earth tones. Many have the hindwing drawn into a tail, which serves to foil the strikes of echolocating bats. In *Copiopteryx*, a favorite of collectors, the tails are three to four times the hindwing length. Perhaps for this same reason, that is to foil bats, the forewing apex is often enlarged and projects outwardly in many arsenurines.

The caterpillar has long, curved or branched, thoracic scoli and a medial horn on A8 through the penultimate instar, which are lost in the final instar, and has a pelage of short, secondary setae, including over the head capsule.

Diets tend to be generalized but with local favorites; many using members of the bombax family (Bombacaceae); also using Annonaceae, Ebenaceae, Lythraceae, Malvaceae, Sapotaceae, and others. *Arsenura* are massive moths—their gregarious larvae, aposematic as early instars but cryptic bark-resters in their final instar, are harvested and eaten by indigenous peoples.

ABOVE | Members of the genus *Copiopteryx* often top the list of "most wanted" when light trapping in the Neotropics.

DISTRIBUTION
New World

IMPORTANT GENERA
Arsenura, Caio, Copiopteryx, Paradaemonia, Rhescyntis, and *Titaea*

HABITAT
Tropical and subtropical forests

HOST ASSOCIATIONS
Very diverse, with both specialists and generalists; temperate genera often on members of oak and pea families

CHARACTERISTICS
• Large moths with comparatively small bodies; wingspans from 2.4–4.0 in (60–100 mm); FW apex often drawn into a point

• Head proportionately small relative to robust thorax and abdomen

• Antennae relatively small

• Caterpillars bear prominent scoli, some often developed into elongate horns

BOMBYCOIDEA: SATURNIIDAE: HEMILEUCINAE
BUCK, IO, AND PANDORA MOTHS

This is the largest saturniid subfamily, with over 630 recognized species referred to more than 50 genera. They are endemic to the New World, with much of their species diversity rooted in the Neotropics; fewer than 40 species are found north of Mexico. *Hemileuca*, a mostly diurnal genus, includes many brightly colored moths.

Hemileucine caterpillars are armed with stinging scoli, which may account for much of the subfamily's evolutionary success. The caterpillars, gaudy in their armament, possess batteries of stinging setae. Many are gregarious as larvae, especially through early instars, a behavior that no doubt links to the effectiveness of their stings as a (collective) defense. Severe stings from the caterpillars of *Lonomia* may lead to internal hemorrhaging and, in extreme cases, fatality. It is unclear if single caterpillars can be lethal—typically the caterpillars rest gregariously on the lower boles of their host when not feeding.

While many hemileucines are somewhat dietarily generalized, by contrast, *Hemileuca* can be highly selective (monophagous) when laying their eggs, but as the larvae mature, they become more generalized in what they will consume. Pandora moths (*Coloradia*) are chronic defoliators of pines in the American West. During population outbreaks, last instars and prepupae were harvested, roasted, and eaten by Paiute tribes. *Hylesia*, which includes the smallest saturniids, are famous for their urticating wing scales; during outbreaks, affected communities will turn out all lights in the hope that no adults will be drawn into the home.

RIGHT | *Coloradia* includes several important defoliators of pines. Native Americans harvested their prepupae from pits dug below infested trees and roasted the caterpillars, which were eaten directly or used in other dishes. Shown here is *Coloradia pandora*.

DISTRIBUTION
New World

IMPORTANT GENERA
Automeris, Coloradia, Dirphia, Gamelia, Hemileuca, Hylesia, Leucanella, Paradirphia, and *Pseudodirphia*

HABITAT
Tropical and temperate forests, woodlands and shrublands, including deserts and savannas

HOST ASSOCIATIONS
Very diverse; most on woody plants; diets range from specialized to highly generalized; some feeding on grasses and cereals

CHARACTERISTICS
• Medium-sized to large moths with wingspans mostly from 2–6 in (50–150 mm); often with large, earth-colored, broadly triangular FW; no FW eyespot; frequently with boldly colored

Hemileuca hera

Hemileuca grotei

Leucanella saturata

Automeris zephyria

THIS PAGE | A selection of various hemileucine adults and caterpillars. Larvae of Hemileucinae are armed with many hundreds of stinging spines. *Lonomia* can be particularly dangerous, as severe stings can lead to internal hemorrhaging.

Cerodirphia candida

HW—several tropical genera (e.g., *Automeris*) have central eyespot on each HW

- Antennae quadripectinate; antennal cones absent
- Frontal protuberance usually present
- Larvae conspicuously armed with stinging scoli; thicker inner portions hollow and poison-filled; distal rami needlelike with sharp points that readily break free from scolus and lodge in skin

Automeris sp.

BOMBYCOIDEA: SPHINGIDAE
SPHINX AND HAWK MOTHS

This well-known group of moths includes over 1,600 species referred to 205 genera. Four subfamilies are recognized: Sphinginae, Langiinae, Macroglossinae, and Smerinthinae, three of which are treated in the following profiles. Sphingidae are the sister group to Saturniidae.

Most are recognizable by their jetlike resting postures, short, rearward-directed wings, and robust bodies, capable of powering rapid long-distance flights and hovering. While their self-powered flight speeds clock in at 12 mph (19 kph), with tail winds some will average speeds over 40 mph (70 kph).

Sphinx moths are reputed to have the most sensitive color vision of all animals—they can discern colors at light intensities that would be perceived as black by other taxa. Their tongues reach incredible lengths, with that of Wallace's Sphinx (*Xanthopan praedicta*) exceeding 11 in (28 cm).

While the majority are nocturnal, diurnality has evolved independently many times across the family. Sphingids are the only bombycoids with

LEFT | *Hyles gallii*, like many other members of the genus, is migratory; it has established itself across boreal lands of the northern hemisphere, and ranks among the few moths that range north to the Arctic Sea.

DISTRIBUTION
Nearly cosmopolitan, with diversity greatest in tropics

IMPORTANT GENERA
Refer to subfamily accounts

HABITAT
Tropical, semitropical, and temperate forests; sclerophyll and shrubland communities; savannas, deserts, and other drylands; many in meadows and open habitats

HOST ASSOCIATIONS
Larval diets tend to be specialized, especially relative to other bombycoids; most feeding on woody seed plants (angiosperms) but others on conifers and nonwoody plants, including some annuals

CHARACTERISTICS
• Wingspans mostly from 1.0–7.5 in (25–190 mm); FW elongate-triangular, often with produced apex and concave outer margin; HW about half length of FW;

bat-detecting organs—the labial palpi of adults have terminal receptors that respond to the high-frequency sounds emitted by bats.

The larvae are the familiar hornworms, with a middorsal scolus that is commonly mistaken as a stinger by other wildlife and people. Their ranks include many important pollinators, especially of tropical and semitropical trees, shrubs, and orchids, as well as a few defoliators and crop pests. Some are used as fishing bait. Many are eaten by indigenous peoples. A few are migratory.

abdomen distinctly tapering rearward in males

- Most have very long tongues but some Smerinthinae nonfeeding, with haustellum reduced
- Male antennae thick through middle, narrowing to apex; never plumose
- Frenulum usually well developed; absent in some Smerinthinae
- Larvae elongate, with middorsal horn on A8, although this may be lost in last instar; often with oblique lateral stripes, with that on A7 continuing to base of horn; sparse, short, secondary setae (usually visible above prolegs)

ABOVE | Three noteworthy hawk moths: (*top*) the diminutive diurnal adults of *Hemaris aethra* are thought to be bumblebee mimics; (*above left*) *Manduca sexta* is a garden pest on tomatoes and lab "rat" for insect physiologists; and (*above*) the stunning *Hyles euphorbiae* has been introduced into Canada as a biocontrol agent to control Leafy Spurge (*Euphorbia esula*).

BOMBYCOIDEA: SPHINGIDAE: SPHINGINAE
SPHINX MOTHS

LEFT | The Greater Death's-head Sphinx (*Acherontia lachesis*) can "talk" when alarmed—the adults squeak by forcing air out of their tongue, and caterpillars make rapid clicks with their mandibles.

This is a modestly diverse subfamily of Sphingidae, with about 275 species in at least 51 genera parsed across 4 tribes. While diversity is centered in the tropics, many genera are well represented in temperate regions.

Adults are strong, powerful fliers, with several migratory species. Included are the longest-tongued moths known, some with tongues that exceed 11 in (28 cm) when fully extended. Sphinx moths are the primary pollinators of hundreds of plants that have tied their fates to these able fliers, capable of seeing under extremely low light conditions and traplining between flowers (like hummingbirds), recalling the locations of nectar-producing flowers in jungles even when individual blooms are well beyond eyesight.

Manduca, the largest genus, with 73 species, includes pests of many members of the nightshade family (Solanaceae). The African Death's-head Sphinx (*Acherontia atropos*) is remarkable in its ability to squeak by pushing air out its pharynx (mouth).

DISTRIBUTION
Cosmopolitan, with diversity greatest in tropics

IMPORTANT GENERA
Acherontia, Agrius, Cocytius, Lintneria, Manduca, Psilogramma, and *Sphinx*

HABITAT
Tropical, semitropical, and temperate forests; various grassland types and drylands

HOST ASSOCIATIONS
Often specialized in diet, particularly on woody plants; many associated with chemically defended plants such as those in bignonia, olive, and nightshade families

CHARACTERISTICS
- Medium-sized to large moths, most with wingspans of 2.7–7.2 in (70–185 mm); FW usually gray or brown but many with mostly black streaks and/or spots; HW usually dark and unpatterned; sometimes brightly colored
- Tongue long to very long
- Caterpillars stout, often with rugose integument, salted with excrescences and transverse creasing

BOMBYCOIDEA: SPHINGIDAE: SMERINTHINAE
EYED SPHINX MOTHS

This is the second-largest subfamily of Sphingidae, with more than 400 species in at least 77 genera. The tribal classification is in flux, with three to six recognized tribes and many genera as yet unplaced. The uncertainty about the number of tribes and tribal membership is a reflection of the great anatomical and biological diversity across the subfamily.

The nominate tribe, Smerinthini, shares many similarities with silk moths (Saturniidae) and other bombycoids: they are bulky and relatively slow-flying, typically have a vestigial haustellum, and eclose with an enormous, egg-packed abdomen. The nonfeeding adults are short-lived and lay nearly all their eggs soon after mating.

By comparison, the Ambulycini are more streamlined and have life histories like other sphingids; that is, they possess long tongues, are active pollinators, can be long-lived, and have much smaller abdomens that continue to mature eggs over days and weeks.

LEFT | The beautiful hindwings of this Blinded Sphinx (*Paonias excaecata*) are concealed when at rest. This moth was alarmed, triggering the flash of pink with embedded eyespots, a common stratagem used by many sphinx and giant silk moths.

DISTRIBUTION
Cosmopolitan, with diversity greatest in tropics and subtropics

IMPORTANT GENERA
Adhemarius, *Ambulyx*, *Polyptychus*, and *Protambulyx*

HABITAT
Tropical, semitropical, and temperate forests; shrublands and chaparral

HOST ASSOCIATIONS
Exceedingly diverse; caterpillars specialists, mostly on broad-leaved trees and shrubs

CHARACTERISTICS
• Large moths; FW elongate with acute apex; strong fliers

• Nonfeeding adults with tongue often reduced or vestigial (e.g., Smerinthini); others with long tongue and avid flower visitors (e.g., Ambulycini and kindred taxa)

• Caterpillars often more elongate than most other sphingids, with small head and cone-like apex; usually with oblique lateral lines

BOMBYCOIDEA: SPHINGIDAE: MACROGLOSSINAE
MACROGLOSSINE SPHINX MOTHS

This is the largest subfamily of Sphingidae, with over 760 species and 88 genera parsed over 3 tribes. It is also the most anatomically and biologically varied subfamily.

Adults range from small, beelike moths to many resplendent moths. Some have ultrasound detectors at the tips of the labial palpi (linked to a cell group at the base of adjacent labral pilifers) to detect hunting bats. Even more extraordinary are those that rub their genitalia together to produce ultrasonic frequencies that disrupt the echolocation signaling of attacking bats and, if successful, avoid capture. Day-flying members of several genera are commonly mistaken for nectaring hummingbirds.

The caterpillars range from cryptic leaf and bark mimics to highly aposematic insects. None can sting with their horn, but they will pretend to do so, thrashing violently, regurgitating, defecating, and biting (harmlessly) when threatened; a few add to the show by hissing or squeaking.

TOP LEFT | Macroglossines are a curious set of comparatively small hawk moths. The day fliers include a number of lineages that are thought to be bee and hummingbird mimics. Shown here is *Cephonodes hylas*.

LEFT | The subfamily includes some of the most skilled fliers that the Lepidoptera can boast. Many are avid nectarers and are collectively thought to be important pollinators, especially across tropical regions.

DISTRIBUTION
Cosmopolitan, with diversity greatest in tropics and subtropics; some genera more speciose in flanking temperate regions

IMPORTANT GENERA
Cephonodes, Eumorpha, Eupanacra, Hippotion, Hyles, Macroglossum, Nephele, Nyceryx, Temnora, Theretra, and *Xylophanes*

HABITAT
Tropical, semitropical, and temperate forests; woodlands, scrublands, meadows, and early successional communities

HOST ASSOCIATIONS
Highly diverse; mostly on broad-leaved woody trees and shrubs, but also herbaceous perennials and annuals

CHARACTERISTICS
• Medium-sized to large moths; many rest with abdomen exposed

• Tongue always present

• Caterpillars diverse; most genera with A8 horn but this lost in some; horn usually without denticles of Sphinginae

GEOMETROIDEA: SEMATURIDAE

AMERICAN SWALLOWTAIL OR CORKSCREW MOTHS

This is a small family with only 40 or so species and 6 genera, but the moths are so beautiful, mysterious, and understudied that they must be mentioned. All but a single African species are endemic to the Neotropics. Sematuridae are related to the Epicopeiidae, swallowtail-like moths endemic to East Asia.

The medium-sized to large adults range from brown and undistinguished to swallowtail-like creatures with patterning continued onto the underwings, often handsomely set with pink, orange, red, and more rarely blue scales. They are wary, rapid fliers. Some fly only at dusk; a few are diurnal. Not many come to lights with consistency, perhaps because many end their activity before nightfall.

The caterpillars are brown bark mimics: they may be smooth or bear raised dorsal protuberances and smaller lateral warting; some are sluglike; and on the whole, they are mundane and rather undistinguished relative to the marvelously colored adults. Because neither the adults nor larvae are commonly encountered, very little is known about the biology and behavior of the family. Here is yet another group especially in need of study.

LEFT | Next to nothing has been published on the adult and larval biologies of sematurids. They are vexingly scarce in collections, perhaps because they are largely dusk-active and only weakly attracted to lights.

DISTRIBUTION
Principally a Neotropical group; one species in southern Africa

IMPORTANT GENERA
Anurapteryx, Apoprogones, Coronidia, Homidiana, and *Mania*

HABITAT
Tropical and semitropical forests

HOST ASSOCIATIONS
Poorly understood; most appear to be specialized in diet, although *Coronidia subpicta* has been reared from eight different plant families

CHARACTERISTICS
• Medium-sized to large; many swallowtail-like with prominent tails; thorax robust
• Antennae curving before acute apex
• Lacks abdominal tympana
• Tongue well developed
• Caterpillars brown and either smooth and undistinguished or with greatly extended dorsal protuberances

GEOMETROIDEA: GEOMETRIDAE
GEOMETRID OR LOOPER MOTHS

LEFT AND INSET | While the vast majority of geometrids are cryptic in coloration and widely palatable, both the adult and larva of *Milionia zonea* shown here are likely unpalatable. Interestingly, the fraction of aposematic geometrids increases as latitudes decrease.

This is the second-largest family of moths, with about 24,000 described species and over 2,050 genera, with recent estimates suggesting that at least 15,000 species have yet to be named. Nine subfamilies are recognized, the four largest of which are introduced in the following profiles.

In habitats with woody vegetation, geometrids are among the most numerically abundant moths and as such play enormous roles in food webs and nutrient cycling. For vertebrate insectivores and especially birds, they are a dietary staple.

While most geometrids are cryptically rendered in earth tones, many unpalatable species are brightly colored. Diurnality has evolved dozens of times across the family—it is common among montane and high-latitude species, as well as among those that are chemically protected, especially in the tropics and subtropics. A significant fraction of the most important outbreak defoliators in temperate forests are geometrids. The adults are often relatively weak fliers and as such are prone to isolation and local endemism. As a

DISTRIBUTION
Cosmopolitan, with diversity highest in tropics

IMPORTANT GENERA
Refer to subfamily accounts

HABITAT
Tropical, semitropical, and temperate forests and woodlands; shrublands and deserts; boreal and montane communities

HOST ASSOCIATIONS
Most on woody, broad-leaved plants (i.e., angiosperms) but also using conifers, ferns, lichens, leaf litter, and debris; one predaceous lineage (*Eupithecia*, in Hawaii) with diet breadths ranging from specialized to broadly polyphagous

CHARACTERISTICS
• Very small to large moths with wingspans mostly from 0.25–2.9 in (6–74 mm); wings proportionately large; body usually slender and approaching butterflies in proportions; females sometimes flightless

consequence, they make good focal taxa for conservation decisions.

The caterpillars, commonly known as inchworms or spanworms, are perplexingly unspecialized in habit: virtually all are external feeders; many are polyphagous, especially those feeding on woody plants; and a few form shelters and nests. What geometrid caterpillars do especially well is hide! They employ a trove of strategies that help them avoid detection: this involves their coloration, shape, texture, resting posture, and use of silk, all of which are a testament to the formidable powers of natural selection, and the pervasive influence of visual predators on the evolution of insects.

ABOVE | *Opisthoxia* are gorgeous moths. Many have wing patterns that direct the observer to (tear-away) eyespots near or along the edge of the hindwings.

BELOW | The *Iridopsis* shown below is a more typical phenotype for the family: bark-like, given to browns and grays, frequently with patterning that continues onto the hindwing.

- Males often with pectinate antennae or otherwise with more obvious chemosensory structures
- Forward-facing abdominal tympana on A2, below juncture with thorax
- Caterpillars variable; most with just one set of midabdominal prolegs on A6; anterior abdominal segments tend to be elongated, while those rearward of A6 tend to be much compressed in length (to facilitate efficiency of looping)

GEOMETROIDEA: GEOMETRIDAE: GEOMETRINAE
EMERALD MOTHS

This is a beautiful group of moths that enjoys special favor among moth collectors and other aficionados. The subfamily includes about 2,700 species and 200 genera. More than 20 tribes have been proposed but the classification of Geometrinae is in great need of a modern revisionary study.

Emeralds are most diverse in the tropics and no more so than in the Neotropics, where many hundreds have yet to be described. Adults tend to be slight, delicate moths, although some Oriental and Indo-Australian genera, which lack green pigments, are more robust in stature. The green pigments responsible for emeralds' ground colors are chemically unique and possibly derive from plant chlorophylls. Relative to the green pigments of most other moths, they are ephemeral, soon fading in bright light and turning yellow when the moths are left in humid containers.

The caterpillars are extraordinary animals: some are elongate; others are stubby with pronounced

LEFT | Geometrines or emeralds have a simple beauty that few would deny: most are green to blue-green, delicate, with pectinate antennae that are tucked under the forewings.

DISTRIBUTION
Cosmopolitan; Neotropics with especially rich fauna

IMPORTANT GENERA
Agathia, Chlorissa, Chlorocoma, Comibaena, Comostola, Dysphania, Eucyclodes, Hemithea, Herochroma, Maxates, Nemoria, Prasinocyma, and *Thalassodes*

HABITAT
Tropical and temperate forests and shrublands, scrublands, and deserts

HOST ASSOCIATIONS
Both host specialists (most) and generalists, mostly on woody, broad-leaved plants (i.e., angiosperms); *Synchlora* and kin often generalized on herbaceous flowers

CHARACTERISTICS
• Small to large moths with wingspans from 0.3–2 in (7–50 mm); ranging from

subdorsal flanges (for example, *Nemoria*). Others still are squat with large subdorsal "wings" that yield a trilobite-like phenotype (for example, *Dichorda*); some comibaenines and nemoriines disguise themselves by silking plant fragments to abdominal protuberances. Geometrine caterpillars are heralded for their phenotypic plasticity: both the body form and color can change depending on what the caterpillars are consuming—such is especially true for the flower feeders.

ABOVE | *Nemoria* caterpillars are among the most phenotypically plastic caterpillars known: *N. arizonaria* caterpillars that feed on oak catkins in spring resemble catkins; summer-generation larvae that feed on leaves resemble twigs; and flower-feeding species are able to take on the colors of the petals they consume.

LEFT | As might be guessed from their color, *Nemoria* are most diverse in forests, woodlands, and shrublands dominated by green vegetation. Many species have both a green and a red-brown morph that is more cryptic when resting on bark.

slim-bodied and delicate to a few robust-bodied genera

• Green pigments chemically unique and relatively unstable

• Two transverse rows of spinelike setae across sternum of A3

• Caterpillars highly variable, from very elongate to short and stubby; integument thick, often with abundant excrescences or minute spines; head frequently cleft to sharply pointed and prothorax drawn into forward-projecting warts or thorns; paraprocts less prominent than in Larentiinae; prominent subdorsal flanges, or "wings," in Nemoriini

GEOMETROIDEA: GEOMETRIDAE: ENNOMINAE

ENNOMINE MOTHS OR THE GRAYS

This is an enormous subfamily with over 11,000 species accounting for almost half of described Geometridae. While cosmopolitan and richly represented in the tropics, much of their diversity is found across temperate regions. Their numerical abundance across many ecosystems makes them one of the most important food sources for the planet's insectivores, with nesting birds being particularly reliant on their ubiquity.

More than 20 tribes have been proposed—the group's higher classification is currently the subject of multiple molecular studies.

The adults are often rendered in grays and browns with darker pattern elements, but yellow, orange, and otherwise brightly colored species abound. They are among the most polyphagous Lepidoptera, especially across temperate and boreal regions. Included are several large-bodied Ennomini that are sometimes confused with owlets (Noctuidae).

Ennomine caterpillars include hundreds of superb background-matchers that mimic leaves,

BELOW LEFT | Due to their sheer abundance, ennomine caterpillars are integral to the diets of many insectivores. Shown here is the Scorched Wing (*Plagodis dolabraria*).

BELOW | Ennomines are masters of crypsis. Many have come to closely match the leaves of their hosts. Pine-feeders have white stripes that mimic the reflections that play off of needles. Shown here is the Tawny-barred Angle (*Macaria liturata*).

DISTRIBUTION
Cosmopolitan, with much diversity found across temperate and boreal forests

IMPORTANT GENERA
Chiasmia, Cleora, Macaria, Pero, and *Zamarada*

HABITAT
Tropical, subtropical, and temperate forests, woodlands, and shrublands; while best represented in mesic habitats, the subfamily is well represented in deserts, chaparral, and other drylands with any woody vegetation

HOST ASSOCIATIONS
Exceedingly diverse, especially on woody plants; diet breadths vary across the 11 tribes, from specialized (e.g., Macariini) to dietary generalists (e.g., Boarmiini)

CHARACTERISTICS
• Small to large moths with wingspans from 0.4–2.9 in (10–74 mm); ranging from slim-bodied to stout-bodied

petioles, twigs, and more, making them essentially invisible to the untrained eye. In many, an individual's phenotype—its shape, color, and patterning—can be greatly influenced by the larva's light environment and diet. The caterpillars are assisted in this matter by their ability to assess their light environment with their skin, as some ennomines have dermal light-sensing capabilities! Included are many orchard and forest defoliators.

More than any other macrolepidopteran group, females have independently (that is, across unrelated lineages) lost their ability to fly and instead rerouted flight resources to elevated fecundity.

The distribution of dark and pale morphs of the Peppered Moth (*Biston betularia*) provides one of the most commonly quoted examples of evolution.

- M2 vein in HW reduced
- Venter of A3 in males frequently with band of rearward-directed setae
- Caterpillars usually cryptic; Boarmiini and others can be exceedingly variable in color; often with protuberances that enhance their crypsis

ABOVE | Many diurnal geometrids are greenish white to bright white (such as this *Campaea perlata*). There is growing evidence that white can represent aposematic coloration in moths, and that many birds ignore white moths.

INSET | Wing veins of moths have tiny amounts of hemolymph moving through them, which means that black scales on the wings of this mountain-dweller (*Chiasmia clathrata*) could play an important role in warming the insect.

GEOMETROIDEA: GEOMETRIDAE: LARENTIINAE

CARPET MOTHS

BELOW | Over 250 species of *Eois* have been named but authorities believe the genus contains at least 1,000 species. Amazingly, nearly all are specialists on *Piper* and *Peperomia* (one species of the former is the source of black pepper).

INSET | In many moths, adult wingspans approximate the fully fed caterpillar's length. The caterpillar shown here (*Eupithecia absinthiata*) and its adult stage wingspan measure about 0.8 in (20 mm).

This cosmopolitan subfamily of geometrids contains some 6,500 species. More than 1,300 of these are *Eupithecia*, also known as the pug moths. A second standout genus, *Eois*, is a pantropical lineage estimated to have more than 1,000 species in the Neotropics alone. Carpets have radiated across temperate and tropical montane regions, with global species richness likely peaking in the mountain ranges of South America.

DISTRIBUTION
Cosmopolitan, with much diversity tied to temperate zones and montane regions of Old and especially New World tropics

IMPORTANT GENERA
Chloroclystis, Chrysolarentia, Entephria, Eois, Eulithis, Eupithecia, Gymnoscelis, Horisme, Hydriomena, Lithostege, Perizoma, Rheumaptera, Scotopteryx, and *Xanthorhoe*

HABITAT
Tropical, subtropical, and temperate forests and woodlands; chaparral and shrublands; many in open meadows; deserts; various montane communities favored worldwide

HOST ASSOCIATIONS
Highly variable, from exceedingly wide-ranging to mostly narrowly specialized on woody and herbaceous plants, even across a single genus

ABOVE | Carpets tend to have complex wing patterns and include many fantastically colored species. This is the Argent and Sable (*Rheumaptera hastata*).

The forewings tend to be handsomely patterned with closely set lines or marbling. While most larentiines feed on woody plants, a fairly large percentage use forbs. Host associations range from broadly polyphagous to highly specialized, with the latter contributing mightily to the subfamily's species richness. As an example, nearly all *Eois* are thought to specialize on just a single or group of closely related *Piper* or *Peperomia* (both Piperaceae) species as caterpillars. Many larentiine caterpillars target flowers and fruits, and some are capable of incorporating floral pigments into their bodies, to better mimic the blossom that they are ingesting.

A very small percentage are serious pests: most famous, perhaps, are the winter moths (*Operophtera*), with their flightless females and males' ability to fly at temperatures at and below 32°F (0°C). A clade of Hawaiian *Eupithecia* are ambush predators that grab and devour flies and other small insects.

(e.g., *Eupithecia*); including conifers, ferns; some predaceous (entomophagous)

CHARACTERISTICS
- Small to large moths with wingspans from 0.3–2.2 in (7–55 mm); often with discal spot in each wing
- HW with vein Sc and R1 merging with Rs near base then separating near middle of discal cell
- Caterpillars slender and elongate (some) to somewhat stubby (many) to grub-like in a few internal borers; usually with distinct paraprocts; often with proportionately small head; integumental excrescences present in many genera

GEOMETROIDEA: GEOMETRIDAE: STERRHINAE
WAVES

This is a speciose subfamily with over 3,000 species and enormous numbers awaiting description. More than 90 percent of the described species are assigned to just three cosmopolitan genera: *Scopula*, *Idaea*, and *Cyclophora*. They and Larentiinae (treated in the preceding profile) separated from other geometrid subfamilies early in the evolution of the family.

The smallest macrolepidopteran moths are sterrhines, with wingspans as small as 0.25 in (6 mm). Contrary to most geometrids, many feed on nonwoody forbs in open habitats; *Idaea* consume living leaf tissue as well as decaying leaves and other organic debris. Many can be lab-reared on clover and other herbaceous vegetation. A few are detritus feeders, with specialized setae, possibly glandular in nature, that enable them to survive as small, exposed, ground-feeding caterpillars where ants, spiders, and other predators abound. The adults of some species of Southeast Asian *Scopula* will feed at blood and eye secretions of vertebrates.

LEFT | Sterrhines include many of the smallest macrolepidopterans—some with wingspans of only 0.25–0.3 in (6–7 mm). Taxonomically they are the most understudied major lineage of macrolepidopterans and offer great opportunities for discovery. Shown here is *Idaea degeneraria*, whose caterpillars are believed to feed on withered and dead vegetation.

DISTRIBUTION
Cosmopolitan, especially through tropics and subtropics

IMPORTANT GENERA
Cyclophora, Idaea, Rhodometra, Scopula, and *Semaeopus*

HABITAT
From forests and shrublands to deserts; some extend into montane and boreal habitats; many associated with open habitats

HOST ASSOCIATIONS
Broad generalists as well as host specialists on both woody plants and herbaceous angiosperms

CHARACTERISTICS
• Very small to medium-sized moths with wingspans from 0.25–1.2 in (6–30 mm); white, gray, or beige with many lines ("waves") that continue from FW onto HW
• Caterpillars slender and elongate (most genera) to short and pudgy (e.g., *Idaea*); some genera with shallow transverse creases ringing abdominal segments

GEOMETROIDEA: URANIIDAE
SCOOPWING AND SUNSET MOTHS

This is a diverse family with almost 800 species and 90 genera divided among 4 subfamilies. The family is sister to the Geometridae.

All share a structurally unique ear on the second abdominal segment, which is tuned to the high ultrasonic frequencies made by hunting bats. Uraniids are anatomically quite diverse, making characterization of the family difficult (separate synopses for two of the subfamilies are included in the following profiles).

While most are brown and other earth tones, and presumably selected to be credible dead-leaf and bark mimics, the family also includes some of the most beautiful terrestrial invertebrates: the diurnal sunset moths are fabulous animals that far surpass the beauty of most butterflies.

BELOW | *Micronia*, endemic to Southeast Asia and Indo-Australia, have lines on the wings that lead to false, tear-away eyespots along the outer hindwing margin (as occurs in *Opisthoxia* geometrids—see page 195).

LEFT | The six swallowtail-like species of *Lyssa* are restricted to Southeast Asia and Indo-Australia. The boldly patterned caterpillars feed on various genera of spurges.

DISTRIBUTION
Cosmopolitan, with much of its diversity tied to tropics

IMPORTANT GENERA
Refer to subfamily accounts

HABITAT
Tropical, semitropical, and temperate forests, woodlands and shrublands; some extending into montane and boreal communities

HOST ASSOCIATIONS
Nearly all specialists on woody, broad-leaved plants (i.e., angiosperms)

CHARACTERISTICS
• Small to large moths with wingspans from 0.6–6.3 in (15–160 mm); often with scalloped wings; posterior margin of HW often projecting to form short to well-developed tail

• Haustellum present, sometimes with prominent sensory setae at base

• Caterpillars cryptic green or brown (most) to boldly colored with flashy spatulate setae (*Chrysiridia*); setae usually borne on warts

GEOMETROIDEA: URANIIDAE: URANIINAE
SUNSET MOTHS

LEFT | Sunset moths such as this *Urania leilus* are spectacular insects. The migration of adults sometimes takes them far out of range.

INSET | Caterpillar of Madagascan Sunset Moth (*Chrysiridia rhipheus*). The larva's black, white, and orange-red (head) coloration warn of its unpalatability.

While the Uraniinae only tally about 50 species and 7 genera, they are marvelous animals of extraordinary beauty. Sunset moths are largely tropical creatures with only the occasional migrant making it into temperate latitudes.

Most are swallowtail-like in habitus, with one or more tails. The three diurnal genera (the sunset moths) are wary, powerful fliers. Included are some of Earth's most beautiful animals: *Alcides* in Australia, *Chrysiridia* in Africa and Madagascar, and *Urania* in the Neotropics. Two of spectacular beauty are the Madagascar Sunset Moth (*Chrysiridia rhipheus*) and Sloane's Urania (*Urania sloanus*, see page 71). The latter went extinct around 1895 due to the conversion of its jungle habitat to agriculture.

The caterpillars are often aposematically colored: boldly marked in black and white with shiny orange-red heads.

The spectacular mass migrations of sunset moths to areas of recent rainfall and newly flushing foliage, sometimes involving many millions of individuals, are one of the most awe-inspiring phenomena of the insect world and further distinguish this wonderful subfamily of moths.

DISTRIBUTION
Cosmopolitan in tropics and subtropics; most diversity in Indo-Australia and Neotropics

IMPORTANT GENERA
Chrysiridia, Cyphura, Lyssa, and *Urania*

HABITAT
Tropical communities; some extending into subtropical and seasonal forests in southern Africa and Australia

HOST ASSOCIATIONS
Specialists on woody spurges (Euphorbiaceae)

CHARACTERISTICS
• Medium-sized to large moths with wingspans from 3–5 in (75–125 mm); robust thorax powering strong flight; HW with one or more tail-like extensions

• Antennae filament-like and rather short

• Ventral surface of tarsomeres 1–4 minutely spined

• Caterpillars stocky; crochets biordinal; *Chrysiridia* and *Urania* bearing elongate setae that are spatulate in some; see also characters for family

GEOMETROIDEA: URANIIDAE: EPIPLEMINAE
SCOOPWING MOTHS

Epipleminae include about three-quarters of the family's diversity, with some 550 species distributed across 70 genera.

Many genera fold or roll the wings such that, at rest, the adults appear to be anything but mothlike. The hindwing often has one or more tails. A few genera have a dark spot proximate to the tail that serves as the eye of a false head, drawing the strikes of predators away from the true head and body; many have lines and other pattern elements on the wings that converge toward the false head to reinforce the deception.

Early instars are often gregarious, living in a silken web but gradually dispersing and becoming solitary as they develop.

BELOW | Most epiplemines can be easily recognized by their unique resting posture with rolled and folded wings, as in this *Phazaca*.

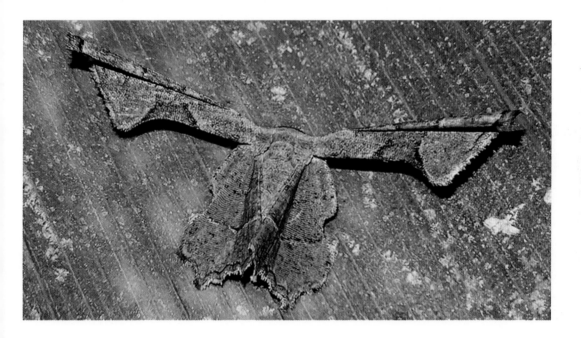

DISTRIBUTION
Cosmopolitan; best represented in tropics and semitropics

IMPORTANT GENERA
Dysaethria, Epiplema, Europlema, and *Phazaca*

HABITAT
Tropical, semitropical, and temperate forests; a few genera extend into boreal plant communities; many tropical taxa inhabiting montane habitats

HOST ASSOCIATIONS
Dietary specialists on broad range of woody, broad-leaved plants

CHARACTERISTICS
• Small to medium-sized moths with wingspans from 0.6–6.3 in (15–160 mm); wings scalloped, well separated at rest

• HW often with tail-like extension of M3 vein; some with additional tails

• Caterpillars with setae usually borne on warts; often stocky in profile

NOCTUOIDEA: NOTODONTIDAE

PROMINENT MOTHS AND PROCESSIONARY CATERPILLARS

The Noctuoidea are the most speciose and arguably most ecologically diverse and important superfamily of Lepidoptera, with close to 45,000 valid described species and perhaps at least half that many again, still unnamed, mostly from the tropics and southern hemisphere. Seven families are currently recognized. The superfamily's extraordinary evolutionary and ecological successes almost certainly tie to their ability to detect the ultrasound emitted by hunting bats—and then evade them.

Approximately 4,500 species and 700 genera of Notodontidae have been described. The family is most diverse in tropical regions, especially the Neotropics, although many subfamilies are well represented in temperate regions, and a few are important elements of boreal and montane faunas. Prominents diverged from other noctuoids 70 mya. More than 20 subfamilies are currently recognized, which alone is indication that the family is both morphologically and biologically diverse, yet all are held together by the possession of a uniquely configured ear, described below.

Notodontid adults tend to be stout and heavyset, drably colored, and night-active. Adults often have resting postures that are twiglike or

LEFT | Prominents are a favored family among those that have a fondness for moths. They have robust bodies, are densely set with hairlike scales, and the thorax often bears scale tufts. At rest, many hold their wings nearly vertically along the sides of the abdomen.

DISTRIBUTION
Cosmopolitan, with exception of New Zealand

IMPORTANT GENERA
Besaia, Hemiceras, Heterocampa, Lyces, Nebulosa, Phalera, Pheosiopsis, Polypoetes, and *Syntypistis*

HABITAT
Tropical, subtropical, and temperate forests and woodlands; chaparral and other shrublands; savannas and deserts, some

extending into boreal and montane communities

HOST ASSOCIATIONS
Exceedingly diverse on woody plants, including monocots, although underrepresented on conifers; important dioptine radiations tied to passion fruit, violet families, and perennial graminoids

CHARACTERISTICS
• Medium-sized to large, heavy-bodied moths; often densely scaled; most drab and

otherwise cryptic, frequently augmented by scale tufts from the thorax or abdomen.

The caterpillars are amazing creatures—fantastically varied in form, color, and behavior. Their tendency to target older leaves does little to speed up their development—a matter that may relate to their marvelous anatomical diversity as a result of protracted periods of larval development and prolonged exposure to birds, monkeys, climbing reptiles, and other visual predators.

TOP | Lobster Caterpillar (*Stauropus alternus*) of southern Asia. Note the extraordinary, spiderlike middle and rear legs; curiously, the first pair of thoracic legs are mere stubs!

SECOND | Adult of *Stauropus alternus*—rather plebian relative to its larva. Its resting posture showing the hindwings is common across many notodontid subfamilies.

THIRD | *Americerura scitiscripta* is a fantastic caterpillar with extensible anal prolegs. In repose, the red and yellow apical bands are withdrawn. When alarmed, hydrostatic pressure extends the warningly colored tissues and the legs are flailed about.

BOTTOM | *Phalera* adults are rather sticklike when settled into their resting posture—and make for credible twigs, a ploy adopted by many notodontid genera.

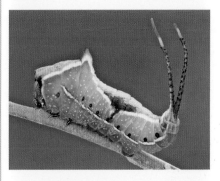

cryptically colored; dioptines often aposematic

• FW and HW with three veins arising from lower end of discal cell

• Metathoracic ear tympanum directed downward and without hood

• Larvae exceptionally varied; anal prolegs often reduced or developed into extrusible tails (stemapods); usually with secondary setae that may be abundant or scarce, sometimes confined to area above prolegs

NOCTUOIDEA: NOLIDAE
TUFTED MOTHS

Nolidae include more than 2,400 species. Much of their species diversity anchors to just two genera: *Nola* and *Meganola*. Nolidae separated from the Noctuidae about 65 mya. They occur worldwide, with greatest diversity in the tropics and subtropics of the Old World; a few genera (for example, *Nycteola*) extend into boreal and montane communities.

While the monophyly of Nolidae is well supported (that is, all trace to a single common ancestor), according to independent molecular studies, no morphological character is shared across all members. Anatomically, they are among the most heterogeneous macrolepidoptera—as an indication of such, more than a dozen different subfamilies have been proposed for this smallish family. Several subfamilies have a tymbal (sound-producing organ) on the abdomen that may play a role in their interactions with bats.

The caterpillars have the normal complement of setae (*Nycteola*) or possess abundant secondary setae (Nolinae). In the latter case, the setae are often

RIGHT | *Paracrama dulcissima* is a particularly handsome nolid from India and Southeast Asia. Males have bright hindwings that become obvious in flight and likely serve as an attention grabber that can be instantly concealed as soon as the animal alights.

DISTRIBUTION
Cosmopolitan; diverse across Old World tropics from Africa to East Asia and Indo-Australia

IMPORTANT GENERA
Afrida, *Characoma*, *Earias*, *Hampsonola*, *Manoba*, *Meganola*, *Nola*, *Nycteola*, and *Risoba*

HABITAT
Tropical and subtropical communities; favoring mesic to wet communities

HOST ASSOCIATIONS
Most specialists on woody broad-leaved plants; some generalists (e.g., *Garella*); forbs, grasses, and other monocots also used; Afridinae are lichen feeders

CHARACTERISTICS
• Very small to medium-sized moths; wings often with gray, white, and black scaling; FW often with raised scale patches (Nolinae and Sarrothripinae); palpi often form projecting snout; quadrifine HW venation

borne from warts called verrucae, as in tiger moths. A few are agricultural pests: *Nola cereella* feeds on sorghum; spiny bollworms (*Earias*) tunnel into and consequently damage developing cotton bolls in the Old World; and the Black-olive Caterpillar (*Garella nilotica*) is a pantropical, polyphagous defoliator. Adults of *Arcyophora* will feed at the eyes of cattle and other mammals.

INSET | Caterpillar of Australia's Gum Leaf Skeletonizer (*Uraba lugens*) with its crown of stacked head capsules from each of the five earlier instars. One wonders if the stacked heads might serve as a club that can be snapped about to knock small enemies away.

- First segment of abdomen with thickened bar behind spiracle
- Larvae often with reduced or missing proleg on A1; many with one SV seta on T2 and T3 and two on A1; commonly with secondary setae borne from verrucae
- Cocoon frequently elongate with vertical slit at one end for egress of adult

ABOVE LEFT | Nolids are exceedingly diverse and difficult to characterize. They tend to have a triangular aspect when viewed from above and a pronounced snout. The nominate subfamily, mostly grayish moths, often bear small tufts on the forewings. Shown here is *Nola cucullatella*.

ABOVE RIGHT | *Sinna calospila*, an exceptionally gaudy nolid relative to the gray adults typical for the family. It is native to montane areas of the Malay Archipelago.

NOCTUOIDEA: NOCTUIDAE
OWLET AND CUTWORM MOTHS

The Noctuidae is the third-largest family of macrolepidopterans, with more than 12,000 species and 1,150 genera. But far more than their species diversity, they have oversized ecological and economic impacts in being among the most abundant macrolepidopterans in many terrestrial communities—their caterpillars frequently outweigh other insect groups in graminoid-dominated communities.

They are diverse in the tropics and temperate zone and are important elements of boreal and montane communities, where many are day-flying. Contrary to most moths, they are much more diverse in temperate biomes than in the tropics, where their importance is replaced by Erebidae.

BELOW | Stiriine noctuids have co-radiated with composites or sunflowers across the deserts and semi-deserts of Mexico and the American Southwest. Their caterpillars are seed predators. Shown here is *Stiria rugifrons*.

BELOW RIGHT | Agaristine noctuids tend to be aposematic both as caterpillars and adults. The caterpillar here (*Eudryas grata*) is quick to vomit on attackers.

DISTRIBUTION
Cosmopolitan; most subfamilies with greatest richness in temperate regions

IMPORTANT GENERA
Refer to subfamily accounts

HABITAT
Virtually all terrestrial habitats in which insects abound; often a dominant group in grasslands, deserts, and early successional communities

HOST ASSOCIATIONS
Virtually all plant taxa and living plant parts eaten; many borers and subterranean species; several lineages specialize on graminoids

CHARACTERISTICS
• Very small (e.g., Eustrotinae) to medium-sized moths; often rendered in earth tones but many brightly colored; usually with stout thorax and capable fliers; many folding wings flat over dorsum

LEFT | The Ear Moth (*Amphipoea oculea*), a widely distributed Palearctic moth, is so named because of the earlike white spot in the forewing, and not for any tendency for it to lodge in human ears—an unsettling experience, I can assure you, that is easily prevented with ear plugs or a bit of cotton tissue when working a light.

Both the adults and larvae of noctuids are difficult to characterize given their anatomical and biological diversity. Most have trifine venation; anatomical details of the ear and male genitalia and associated musculature hold the family together. While most are drab and cryptic in coloration, the family includes dozens of extremely handsome lineages. The same might be said of their larvae, which range from the most mundane cutworms to gorgeous agaristine and cuculliine caterpillars.

Many serious crop pests are migratory noctuids that move from areas with mild winters into temperate latitudes in spring and summer. Propelled by tropical storm fronts, the adults can pour into a region by the tens of millions overnight—these are the bollworms, cutworms, earworms, and armyworms that have plagued humankind since the dawn of agriculture. Yet, noctuids likely pollinate more flowers than any other group of Lepidoptera.

- HW usually with three veins that arise from lower end of discal cell (trifine condition)
- Haustellum well developed
- Metathoracic ear directed ventroposteriorly with dorsal hood; anterior of abdomen modified into counter-tymbal to reflect sound back to tympanum
- Larvae exceedingly varied; early diverging lineages missing first two pairs of prolegs or with first two pairs of prolegs somewhat reduced (especially in first instar); commonly only two SV setae on A1

NOCTUOIDEA: NOCTUIDAE: NOCTUINAE
DART, ARMYWORM, AND CUTWORM MOTHS

BELOW | Pinions, mundane in coloration, make for challenging quarry. All are "winter moths" that overwinter as adults in leaves and woodpiles. Many are infrequent at lights and seen more reliably at sugary (fermenting) baits. Shown here is *Lithophane leautieri*.

INSET | Virtually all of the 90 known species of pinions have caterpillars that blend in well with their host. As could be guessed from its coloration, *Lithophane leautieri* feeds on cedars and related conifers.

In North America and Europe, this subfamily accounts for more than 55 percent of described Noctuidae. Much of the subfamily's species richness, ecological function, and economic importance anchors to just two tribes: the cutworms (Noctuini) and armyworms and kin (Hadenini).

The caterpillars of cutworms leave their host plants by day to hide under litter or tunnel into soil—a matter that has likely played a role in their

DISTRIBUTION
Cosmopolitan; pest taxa ranging into temperate croplands of both hemispheres

IMPORTANT GENERA
Acrapex, *Agrotis*, *Apamea*, *Athetis*, *Caradrina*, *Dichagyris*, *Elaphria*, *Eriopyga*, *Euxoa*, *Hadena*, *Leucania*, *Mythimna*, and *Spodoptera*

HABITAT
Many plant communities but richest in open grasslands, meadows, wetlands, early successional habitats, and other communities dominated by forbs and grasses; xylenines abound in temperate forests, woodlands, and shrublands

HOST ASSOCIATIONS
Often generalized on low plants, both woody and herbaceous, including graminoids; by contrast, xylenine noctuids tend to feed on woody plants and may have specialized diets

successes. Relative to most moths, noctuines have flourished on herbs, grasses, and other graminoids, and included among the Noctuinae are many of the world's most damaging crop pests. The Army Cutworm (*Euxoa auxiliaris*) and Bogong Moth (*Agrotis infusa*) are featured in the Introduction (pages 54–55).

Also included here are the "winter moths" that emerge in the fall and overwinter either as eggs (for example, *Eupsilia*) or as prereproductive adults (for example, *Lithophane* and *Xylena*). These and other xylenine noctuines account for the lion's share of larger moths seen in headlights from October to March through north temperate regions, capable of freezing and thawing repeatedly through the winter months. On nights when temperatures rise, they will feed at broken limbs, sap fluxes, leftover fruits, and maple syrup buckets (where they are most unwelcome), as well as at the artificial sugar baits used by collectors, in good years arriving by the hundreds to feed on (often fermenting) sugary solutions.

TOP | Quakers (*Orthosia*) are among the most abundant caterpillars of woodlands and forests across the northern hemisphere in the spring, and as such are a food staple for many songbirds.

MIDDLE | Army Cutworm (*Euxoa auxiliaris*) adult, perhaps on its way to seek out subalpine scree where it can aestivate over the summer.

BOTTOM | The Bogong Moth (*Agrotis infusa*) aestivated by the billions in caves and rock crevices in the mountains of southwestern Australia, where it was feasted upon by indigenous peoples and countless insectivores. But no more—see page 42.

CHARACTERISTICS
- Small to medium-large moths; often rendered in gray, brown, and other earth tones; FW elongate-triangular with truncate outer margin; many (darts) fold their wings flat over the abdomen
- Caterpillars commonly with SD1 seta on A9 thin, serving in mechanoreception (e.g., in sound detection)
- Larval spinneret often short, flattened, and apically flared

NOCTUOIDEA: NOCTUIDAE: PLUSIINAE
SEMI-LOOPER MOTHS

LEFT | Can't live with them; can't live without them. Plusiines include some of the world's worst crop pests. Yet, their adults are important pollinators, especially in montane and boreal communities. Shown here is *Autographa precationis*.

BELOW | Semi-loopers like this *Autographa precationis* commonly are missing the first two pairs of prolegs, and as a consequence loop their bodies like geometrids when moving between perches.

About 500 plusiine species are recognized. The 50 or so genera are grouped into 3 tribes: the mostly tropical Abrostolini and Argyrogrammatini and the temperate and boreal Plusiini.

Adults are avid nectarers and likely undervalued pollinators, especially in subpolar, alpine, and mountain habitats, where other insect diversity can be modest. Many become active at dusk and are fun to watch working flowers—if you see a noctuid placing both its tongue and front legs into a corolla throat, while still in flight, you are likely seeing a plusiine. In colder environments and during their migrations, adults will feed by day.

Plusiines include more than two dozen of the world's most damaging lepidopteran crop and garden pests, with most of these belonging to *Autographa*, *Chrysodeixis*, *Rachiplusia*, *Thysanoplusia*, and *Trichoplusia*—all of which are well-known migrants. Caterpillars of most loop when they move (hence their vernacular name of semi-loopers), in a fashion similar to inchworms (Geometridae), as most lack prolegs on A3 and A4.

DISTRIBUTION
Cosmopolitan; most diverse through temperate areas

IMPORTANT GENERA
Autographa, *Chrysodeixis*, *Ctenoplusia*, *Syngrapha*, and *Thysanoplusia*

HABITAT
Open communities: high-latitude and montane meadows, heathlands, bogs, and other open, mesic habitats; early successional habitats, including croplands

HOST ASSOCIATIONS
Low plants, both woody and herbaceous, including graminoids

CHARACTERISTICS
• Medium-sized, brown to gray moths, commonly with lustrous silvery or brassy scaling

• Dorsum of thorax often with tuft(s) of scales

• Adults tend to have long tongues

• Caterpillars usually lime green and missing first two pairs of prolegs (Plusiini)

NOCTUOIDEA: NOCTUIDAE: HELIOTHINAE
FLOWER MOTHS

This is a smallish subfamily with only about 400 described species but includes many globally important crop pests (*Helicoverpa*, *Heliothis*, and *Heliocheilus*) and a fascinating radiation of western North American moths (*Schinia*) that enjoys much attention from collectors, photographers, and conservation biologists.

While many heliothine caterpillars feed on leaves, most target flowers and especially young fruits and the seeds therein and will die if fed only leaves. Pest species of *Helicoverpa* and *Heliothis* tend to be broadly polyphagous—and are chronic crop pests, inflicting devastating damages to grain, cotton, vegetable, and other field crops. The Cotton Bollworm (*Helicoverpa armigera*), an Old World species, is responsible for major losses annually. By contrast, *Schinia* tend to be highly specialized—some only nectar at their host species and can occasionally be observed ovipositing and nectaring simultaneously!

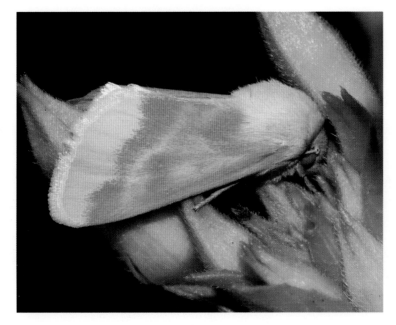

LEFT | A Primrose Moth (*Schinia florida*) resting on its larval host (*Oenothera*). More typically, adults will rest tucked into the flowers. *Heliolonche* adults perch in host flowers that close at night, affording them a safe night's stay.

DISTRIBUTION
Cosmopolitan; pest genera diverse and numerically abundant primarily through tropics and subtropics

IMPORTANT GENERA
Helicoverpa, *Heliothis*, and *Schinia*

HABITAT
Many plant communities but richest in deserts and other drylands of North America

HOST ASSOCIATIONS
Low plants, both woody and herbaceous, including graminoids; ranging from exceedingly generalized to specialized on single host-plant species

CHARACTERISTICS
• Small to medium-sized moths; often rendered in earth tones but many brightly colored, especially among diurnal species

• Protibiae often with clawlike setae

• Larval integument densely set with minute spinules (visible with lens); L1 and L2 setae are aligned horizontally on prothorax

NOCTUOIDEA: EREBIDAE
EREBID AND UNDERWING MOTHS

BELOW | Erebids are exceedingly diverse in species, appearance, and ecology. Adults are often triangular in dorsal view with a short snout (and metathoracic ear). Many have bark-like forewing patterns, as seen here in *Metalectra diabolica*.

INSET | Late instar *Metalectra* caterpillar, which specializes on fungal hyphae. The caterpillars tunnel through mushrooms and graze on fungal mats, usually secreted away in tree holes and under bark and rotting logs.

Erebidae are the most species-rich family of moths, with over 25,000 recognized species and new species, genera, and even tribes being discovered at a fast rate. Erebidae separated from their kindred noctuoids (Nolidae, Euteliidae, and Noctuidae) about 60 mya. As is commonly the case with taxonomically and ecologically diverse insect families, no single character unites the members of the family.

DISTRIBUTION
Cosmopolitan, with highest species diversity in tropics and subtropics; hyperdiverse in Neotropics

IMPORTANT GENERA
Refer to subfamily accounts

HABITAT
All terrestrial habitats; dominant group across tropical and subtropical communities but especially forests and woodlands

HOST ASSOCIATIONS
Very diverse: most are host specialists on woody plants and require new growth to complete development; conifers, forbs, grasses and other monocots, and ferns favored by some; other lineages target algae, lichens, and fungi; Herminiinae have radiated on fallen leaves and other detritus; arctiines have diversified on many plants too toxic for most other herbivores

The adults tend to be more slender than most noctuoids and have broader wings. It is not obvious why there are so many more erebids. Part of the explanation anchors to the Arctiinae, or tiger moths, which account for more than 11,000 species. They are extraordinary animals, worthy of their own series of books—they and four other subfamilies are treated separately in the following profiles.

Some aspects of erebid ecology and behavior differ somewhat from those of other noctuoids, which may relate to their evolutionary successes. The adults tend to feed at sap, tree wounds, fermenting fruit, and honeydew rather than at flowers. Many are migratory.

The larvae are exceedingly diverse: commonly they are elongate with some reduction of the anterior prolegs, which allows them to move more efficiently. Many are new-leaf specialists that commute great distances on any given night to find young leaves. In the Erebinae, which is one of the largest subfamilies, the gray and brown erebid caterpillars abandon the leaves by day to rest on twigs, branches, and stems, only to return after nightfall.

LEFT | *Eulepidotis*, with nearly 100 species across the Neotropics, shares a reoccurring stratagem adopted by many moth lineages, that is they have forewing markings that lead to a tear-away, false-head, pattern along the trailing edge of the hindwings where a bird strike would do the least damage.

CHARACTERISTICS
- Very small to largest moths; often rendered in black, grays, whites, and spectrum of earth tones; HWs, which are hidden at rest, sometimes brightly colored; body more slender than many noctuids; FW pattern commonly continued onto HW
- M2 vein in HW arising adjacent to M3 such that four veins arise from lower end of discal cell (quadrifine condition)
- Haustellum well developed; lower part of face unscaled; labial palpi often upcurved
- Caterpillars exceedingly varied; many elongate, muscular, with reduced anterior prolegs; often gray or brown bark-mimicking larvae that leave foliage by day

NOCTUOIDEA: EREBIDAE: HERMINIINAE
LITTER MOTHS

This is a large subfamily, but a global compilation of species has not yet been made.

They are among the more poorly studied erebids due to their small to modest size, drab coloration, and enormous diversity. They are richly represented in forested ecosystems, including temperate woodlands, marshes, and other mesic environments with abundant plant litter.

Male herminiines have a splendid array of scent (androconial) brushes—on the antennae, palps, thorax, legs, wings, and sternite on A8. Adults feed at sugary substrates and generally shun flowers.

Larvae consume a huge range of organic matter, including living foliage of flowering plants, conifers, and ferns; algae and lichens; fallen leaves, flowers, and fruits in various states of decomposition; and some consume dead insects, fungi, and vertebrate dung. Their ecological importance as macrodecomposers, life histories, and ecological function in food webs are much in need of study.

LEFT | Herminiines are unsung heroes in many forests, playing important roles as macrodecomposers of leaves and other debris. Adults feed primarily at plant exudates and other nonfloral resources.

DISTRIBUTION
Cosmopolitan

IMPORTANT GENERA
Bertula, Bleptina, Idia, Hydrillodes, Naarda, Renia, Simplicia, and *Zanclognatha*

HABITAT
Tropical, subtropical, and temperate forests and woodlands; wetlands; mesic habitats with abundant plant litter

HOST ASSOCIATIONS
Many aboveground leaf feeders but most on forest-floor leaves, organic litter, and undergrowth

CHARACTERISTICS
- Small to medium-sized moths with wingspans from 0.5–1.4 in (12–35 mm); mostly brown or gray, relatively nondescript
- Long upcurved palps that may extend above and over head; unscaled lower frons
- Larvae earth-colored with roughened integument; short, inconspicuous, and often peg-like setae; prolegs on A3 sometimes reduced or absent

NOCTUOIDEA: EREBIDAE: LYMANTRIINAE
TUSSOCK AND SATIN MOTHS

This is a large subfamily of erebids with more than 2,900 species parsed across 360 genera. The adults are nonfeeding, with vestigial mouthparts. And while the adults are lacking digestive capabilities, they are renowned for their reproductive résumés. Male lymantriines have proportionately enormous antennae and preeminent abilities to detect and navigate upwind to infinitesimal concentrations of the female sex pheromone. Females excel in egg production, with abdomens so packed with ova that many can't fly until a first clutch of eggs is laid. And in the extreme, some female lymantriines have foregone flight

ABOVE | Northerly populations of the Arctic Woolly Bear (*Gynaephora groenlandica*) occur north of 80° latitude.

INSET | DO NOT HANDLE: The Brown-tail Moth (*Euproctis chrysorrhoea*) caterpillar possesses batteries of urticating, deciduous setae along its dorsum that cause serious dermal reactions that may last for a week or more.

DISTRIBUTION
Cosmopolitan; most diverse in Old World tropics, especially across Africa, India, and Southeast Asia

IMPORTANT GENERA
Arctornis, Euproctis, Laelia, Leptocneria, Leucoma, Lymantria, Orgyia, and *Rhypopteryx*

HABITAT
Forests, woodlands, and shrublands

HOST ASSOCIATIONS
Most temperate species are polyphagous on woody plants, including broad-leaved plants (angiosperms) and conifers; a few genera feed on forbs and grasses; most surprising, some appear to develop on leaf litter

entirely, sporting foreshortened or vestigial wings and tiny thoraces incapable of powering flight.

The caterpillars of *Orgyia*, *Dasychira*, and kin have dense dorsal tufts of setae (tussocks) over A1–A4. *Euproctis*—an enormous genus with over 500 species—may have short, hollow, toxin-laden setae that trigger severe dermatitis and allergic reactions. Prepupal lymantriines frequently weave their setae into the walls of their cocoons for protection; and in some taxa, particularly those with flightless females, these same setae will be gathered by the female and pulled into the foamy covering that she places over her egg mass.

Dispersal in lymantriines with flightless females is achieved by ballooning—neonates and early instars drop on a silken thread to be carried off by winds to new locations.

BELOW | Male *Lymantria dispar* at the ready. Each antennal branch is lined with hundreds of sensilla specialized to detect the female sex pheromone.

CHARACTERISTICS
- Medium-sized moths, mostly drably colored and nocturnal; some tropical species white with satiny sheen; FWs often rounded-triangular, held back and flattened against substrate with hairy forelegs extended forward

- Broadly plumose male antennae held erect

- First abdominal segment with counter-tymbal hood anterior to spiracle

- Caterpillars abundantly vested in secondary setae borne from warts (verrucae); dorsal setae often gathered into dorsal tufts (tussocks); brightly colored middorsal defensive gland on A6 and (usually) A7

SPONGY MOTH

LEFT | Last instar of Spongy Moth. Female caterpillars often go through an extra instar and attain greater lengths and mass that will be converted into eggs in the pupal stage.

Native to the Palearctic region, the Spongy Moth (*Lymantria dispar*), formerly the Gypsy Moth, was introduced into North America in the 1860s as a result of a failed attempt by Étienne Léopold Trouvelot to make commercial silk from the cocoon. Once established, it became a chronic pest, whose range is still expanding westward across forests in eastern North America and southern Canada. In the US alone, *L. dispar* defoliates 700,000 acres each year, causing more than US$200 million in annual damages. Nearly 13 million acres of forest were defoliated in 1981 across North America.

The caterpillars are especially damaging to oaks, frequently killing older trees over the course of multiyear outbreaks or when defoliation is accompanied by droughty conditions, at great cost to homeowners whose dead trees must be cut down and removed.

In the Old World, where the species is native, outbreaks are reduced in frequency, extent, and duration, although east Asian races, which often favor conifers, are chronic forest pests across eastern Russia. There is great concern that the introduction of Eurasian conifer-feeding races to the western US and Canada would be devastating to Douglas Fir and other important lumber trees of the Pacific Northwest and western Canada.

The caterpillars are protected by histamines and should be handled gently or not at all; those that are handled carelessly or find their way into clothing may elicit severe allergic reactions in sensitive individuals.

The adult males are notorious for their ability to track tiny traces of the female sex pheromone. Scientists who work in Spongy Moth labs and have absorbed infinitesimal titers of the pheromone in their body tissues (and urine) have had excited males show up at their weekend hikes and picnics—and at outdoor weddings, as especially unwelcome guests!

NOCTUOIDEA: EREBIDAE: ARCTIINAE
TIGER MOTHS

Tiger moths are a crowning achievement of the Lepidoptera and, more broadly, of biological evolution. They are hyperdiverse, with more than 11,000 species. Arctiines account for much of the aboveground insect biomass across many tropical and subtropical ecosystems and thus have an oversized impact on ecosystem function in the richest terrestrial biomes on Earth.

Four tribes are recognized: Amerilini, Arctiini, Lithosiini, and Syntomini, the three latter of which are introduced separately below. The subfamily arose between 25 and 50 mya, at about the time bats began to flourish ecologically—their storied abilities to thwart bat predation almost certainly link to their extraordinary evolutionary and ecological successes. Their unrivaled defense pharmacopoeia ranges from simple, self-manufactured bioamines, histamines, and pyrazines to highly toxic and structurally complex

BELOW | Tiger moths tend to be boldly patterned, to warn of their toxicity or to mimic unpalatable kindred. The Faithful Beauty (*Composia fidelissima*) is nothing short of gaudy.

DISTRIBUTION
Cosmopolitan; most diverse in tropics; hyperdiverse in Neotropics

IMPORTANT GENERA
Refer to tribal accounts

HABITAT
Tropical and temperate forests, woodlands, and shrublands; present but less diverse in open communities dominated by forbs and grasses

HOST ASSOCIATIONS
Exceedingly diverse; consuming essentially all plant taxa, as well as lichens and algae; many feed on plants too poisonous for other Lepidoptera

CHARACTERISTICS
• Small to medium-sized moths, most with bright or bold colors that warn of their defensive chemistry

• Adult females have paired, eversible, pheromone-emitting glands on A8

compounds sequestered from their host plants; for example, cardenolides, iridoid glycosides, pyrrolizidine alkaloids, and lichen phenolics. Accordingly, both the adults and larvae are often brightly colored, warning of their unpalatability.

Arctiine larvae are renowned for their ability to consume plants that are toxic to most insects, ungulates, cervids, rabbits, and other herbivores, thanks to the batteries of detoxifying enzymes housed in the caterpillars' fat body. The caterpillars add a new dimension to the generalist-to-specialist diet axis used to characterize most moth families in this guide; many arctiines, especially the woolly bears and other ground-dwelling taxa, are diet mixers that vary their diets over the course of a day or night, consuming leaves of many different plant species on their walkabouts.

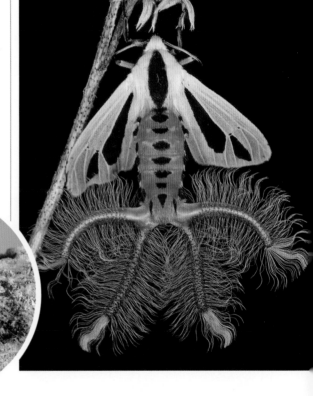

- Tymbal (i.e., sound-producing) organ under the metathoracic ear, although non-functional in some taxa
- Caterpillars nearly always generously vested in minutely barbed secondary setae; D1 and D2 setae borne from common wart (verruca) on T2 and T3; crochets in series roughly parallel to body axis with anterior and posterior crochets much reduced in size

INSET | Larva of lichen moth—a splendid radiation of nearly 3,000 species anchored to the tropics but extending to the High Arctic. While many consume lichens, it is largely the algal component that is essential.

TOP | The diversity and relative abundance of arctiines increase as one approaches the equator. In light trap collections from the Neotropics, tiger moths often account for much of a sample's biomass. Shown here is *Vamuna remelana*.

ABOVE | Sexual selection often drives animal anatomy and behavior to seemingly ridiculous extremes. Here, a male *Creatonotos gangis* is deploying his pheromone-emitting coremata—and these, still, are only half inflated!

NOCTUOIDEA: EREBIDAE: ARCTIINAE: ARCTIINI
TIGER MOTHS

This taxon includes some of Earth's most extraordinary insects. They have species diversity, endless beauty, fascinating host-plant interrelationships that underlie much of their natural history, extraordinary courtship and mating behaviors, and have been embroiled in intense arms races with bats for more than 30 million years.

About 7,000 species have been described. They are most ecologically and evolutionarily successful in the Neotropics. At sites in the eastern foothills of the Andes, more than 350 species may occur sympatrically. In light trap samples from the tropics, arctiines will account for more insect biomass than all but the giant silk (Saturniidae) and sphinx (Sphingidae) moths.

Their array of self-manufactured and sequestered defensive chemicals may be unrivaled by any other animal group (see previous profile for Arctiinae). Host associations range from highly specialized to some of the most generalized recorded for any caterpillars; the ground-dwelling woolly bears are particularly catholic in diet.

LEFT AND INSET | At rest, the bright hindwings of the Garden Tiger (*Arctia caja*) are concealed under the closed forewings. If disturbed, the moth "flashes" the hindwings. It is both chemically and physically protected. Caterpillars have long setae to exclude small enemies (the setae cause hives in some humans).

DISTRIBUTION
Cosmopolitan, from tropical regions to high latitudes

IMPORTANT GENERA
Amastus, Hypercompe, Leucanopsis, Lophocampa, Spilarctia, Trichromia, Utetheisa, and *Virbia*

HABITAT
Forests and woodlands; open communities of savannas, shrublands, grasslands, deserts, and open, early successional habitats; a few aquatic as caterpillars

HOST ASSOCIATIONS
Exceedingly varied, including many highly toxic plants shunned by other Lepidoptera

CHARACTERISTICS
• Mostly medium-sized moths with wingspans generally from 0.7–2.5 in (18–64 mm); typically boldly patterned to advertise their unpalatability

• Frenulum well developed in both sexes

• Most caterpillars densely vested in barbed, secondary setae; see also characters given for Arctiinae

RATTLEBOX MOTHS

LEFT AND INSET | The bright coloration of the Bella Moth (*Utetheisa ornatrix*) warns of its unpalatability—the egg, caterpillar, pupa, and adult are all fortified with crotaline alkaloids. When threatened (*inset*), it (and some other tiger moths) auto-hemorrhages from weak areas of the cuticle adjacent to a thoracic trachea that releases air into the leaking hemolymph, forming a sizeable, toxic, frothy bubble to ward off would-be predators.

Rattlebox moths (*Utetheisa*) are a pantropical genus of brightly colored Arctinii. The elegantly colored adults—white to orange or red with bold spotting that warns of their dangerously toxic chemistry—are active both day and night. The caterpillars, too, are boldly marked, usually banded in bright orange and black; like the adults, the caterpillars can be active through the daytime.

Utetheisa feed on species of rattlebox (*Crotalaria*), a plant that is highly toxic to vertebrates and most phytophagous insects, as its tissues and, especially, its seeds are richly protected by pyrrolizidine alkaloids (PAs). Once ingested by vertebrates, attempts by liver tissues to metabolically detoxify the PAs instead render them into pyrroles, even more poisonous chemical products that damage vertebrate liver tissues. Chronic or acute exposure to PAs can lead to animal wasting and deaths in horses, cattle, and others.

The caterpillars, on the other hand, actually retain and accumulate high titers of crotalines and other PAs and carry these through subsequent instars, the pupa, and into the adult stage! It doesn't stop there: males load derivatives of these alkaloids onto the courtship brushes that are deployed during courtship; in this way, a male's titer of the alkaloids can be assessed by the female as a means of evaluating the quality of her suitor. Further, the male's PAs—sequestered when he was a caterpillar—are loaded into his spermatophore (see pages 49–50) and delivered into the female's bursa copulatrix during hours-long mating. The female will, in turn, pass the gained alkaloids into her eggs and in so doing provide chemical protection for the couple's offspring. Low levels of the male's PAs can persist even through the early and middle instars, essentially at the time when the next-generation of caterpillars can start sequestering their own PAs from *Crotalaria*.

Sexually transmitted impunity: male *Utetheisa* transfer as much as 5–10 percent of their body mass to their female partners during mating; that is, with the transfer of their nuptial gift, the spermatophore, which, in addition to sperm and PAs, contains proteins, lipids, salts, and other nutrients. Female *Utetheisa* that have been denied access to PAs as larvae will be greedily consumed by *Nephila* spiders. But once a PA-free female copulates with a PA-laden male, the female *Utetheisa* may be chemically protected from a lethal attack by a spider. *Nephila* spiders will even go so far as to cut and remove PA-containing male and female *Utetheisa* moths from their webs.

NOCTUOIDEA: EREBIDAE: ARCTIINAE: LITHOSIINI
LICHEN OR FOOTMAN MOTHS

This splendid group of diminutive tiger moths includes some 4,000 described species and 350 genera, with many more yet to be described. The tropics are especially rich in lithosiines—with single sites sometimes supporting more than 50 sympatric species.

The aposematic adults are often members of Müllerian mimicry complexes. Several mimic highly poisonous beetles, especially net-winged beetles (Lycidae) and fireflies (Lampyridae); larger lithosiines often mimic wasps, especially in flight.

The caterpillars of some are known to sequester complex phenolic compounds from their host lichens. Given their bold coloration, it is likely that many lithosiines will be found to sequester a smorgasbord of other secondary chemicals from their lichen hosts.

As in other arctiines, adults possess a tymbal that can produce bursts of ultrasonic frequencies to advertise their unpalatability to hunting bats. Lithosiini represent a wonderful group of greatly understudied moths.

LEFT | *Cyana* is an enormous Old World genus with nearly 200 species. All but a handful are mostly white and boldly marked in red, orange, yellow, and black, presumably advertising their unpalatability. The defensive chemistry of lithosiines is poorly understood. Shown here is *Cyana malayensis*.

DISTRIBUTION
Cosmopolitan; most diverse in Neotropics

IMPORTANT GENERA
Aemene, Agylla, Ammatho, Cisthene, Cyana, Eilema, and *Eugoa*

HABITAT
Tropical and temperate forests and woodlands; shrublands, savannas, and scrublands; modest representation elsewhere

HOST ASSOCIATIONS
Lichens and sometimes terrestrial algal layers

CHARACTERISTICS
- Small (most) to medium-sized moths with wingspans from 0.5–0.8 in (13–21 mm), exceptionally to 1.4 in (35 mm); typically with bold red, orange, or yellow markings on FWs, HWs, and/or abdomen
- Ocelli highly reduced
- Larval mandible with central grinding tooth (mola) on inner surface
- Most larvae with dense, minutely barbed secondary setae; some lack secondary setae

NOCTUOIDEA: EREBIDAE: ARCTIINAE: SYNTOMINI
WASP AND HANDMAIDEN MOTHS

Syntomini include about 800 species and 80 genera anchored to the Old World tropics and subtropics, although Madagascar and much of Africa have a rich fauna. Many are boldly patterned and day-flying—traits associated with being chemically protected. It would be expected that almost all are actors in mimicry complexes common to other syntomines, more distantly related tiger moths, smoky moths (Zygaenidae), and various wasps. However, the nature of their defensive chemistry remains puzzling given many are broadly polyphagous on low-lying plants, and sometimes leaf litter. Such catholic diets are rare among chemically protected species—most aposematic day-flying moths are dietary specialists that sequester defensive chemicals from their (toxic) host plants.

ABOVE | There are over 100 *Amata*, all found in the Old World. While many are mimetic of wasps and other moths, it is not clear to what degree they are Batesian or Müllerian mimics. Shown here is *Amata compta*.

DISTRIBUTION
Old World through tropical regions; diversity dropping off rapidly with latitude

IMPORTANT GENERA
Amata, Balacra, Ceryx, Thyrosticta

HABITAT
Tropical communities, both forested and especially open communities

HOST ASSOCIATIONS
Diverse, including broad-leaved plants, woody trees, shrubs, and herbaceous plants; a few on grasses

CHARACTERISTICS
• Wingspans range from 0.6–1.8 in (15–45 mm); elongate FW and small HW

• Wasplike; many black with white spots; commonly with clear areas on wings

• HW veins Sc + R and Rs fused; vein 2A in HW usually absent

• Ventral pheromone gland

• Larvae range from dark ground dwellers to boldly rendered in orange and black; crochets essentially of single length

NOCTUOIDEA: EREBIDAE: EREBINAE
UNDERWING MOTHS AND WITCHES

The taxonomic breadth of Erebinae, the nominate subfamily, has been greatly expanded over the past two decades as molecular data have helped elucidate the phylogenetic history and taxonomic bounds of the group. Current concepts of the subfamily include some 2,500 described species, most of which were formerly classified as "catocalines." Anatomically and ecologically, they are diverse in form and habitat, such that only a coarse characterization of this subfamily can be provided here.

While the forewings tend to be rendered in bark-like or other earthen colors to aid in crypsis, the hindwings are sometimes given to bright yellow, orange, or red colors that can be flashed when the moth is threatened by a would-be enemy. Such brightly colored species of Catocalini, Melipotini, and Ophiusini are among the most sought-after noctuoids. Rather than nectar, the adults of some lineages favor nonfloral resources, feeding at fermenting fruits, tree wounds, and honeydew—reflecting the fact that many are forest moths that can get along without the flowers required by most Lepidoptera.

Both adults and caterpillars often perch on branches and tree trunks by day. Most are host-plant specialists, with a great proportion tied to new foliage rich in water and nutrients, before the leaf tissues become protected by the defensive chemicals common to older foliage. Many are strong fliers that migrate long distances to ensure access to new foliage for their offspring. Grass-feeding lineages include pest species that sometimes decimate sugarcane, cereal crops, and pasturelands. Fruit-piercing Ophiderini can be important pests of orchard crops. *Melipotis* and kin are occasional defoliators of their mimosid host plants.

DISTRIBUTION
Cosmopolitan, with greatest diversity across tropics; Neotropics especially rich

IMPORTANT GENERA
Catocala, Dysgonia, Erebus, Metria, Mocis, and *Ophiusa*

HABITAT
Tropical and temperate forests, woodlands, and shrublands; also savannas and grasslands, and some dryland communities

HOST ASSOCIATIONS
Most specialists on woody plants, including broad-leaved plants (angiosperms) and conifers; large fraction using legumes; some grass-feeding lineages

CHARACTERISTICS
• Small to very large moths with wingspans from 0.5–12 in (12 mm–30 cm); FWs usually with black, gray, and brown bark-like patterning that may continue onto HWs; often resting with HWs partially

OPPOSITE | Over 190 species of *Catocala*, a perennial favorite with collectors and moth photographers, are recognized. The Holarctic genus is especially diverse across North America, boasting 110 species. Shown here is a Clifden Nonpareil (*Catocala fraxini*).

visible; those with boldly colored HWs hide these at rest

• Well-developed haustellum with smoothly tapering apex

• Seventh sternite of females deeply incised with ostium bursae (mating orifice) shifted anteriorly into cleft

• Caterpillars elongate, tapered rearward, often with reduced prolegs on A3 and A4 and paired warts over A8 bearing D2 setae

TOP | Like many erebids, *Melipotis* are migratory, moving from areas of high density to new woodlands with abundant new foliage. Several are important defoliators. By day the late instars will roost off-plant and under bark, then erupt under cover of darkness to feed. Shown here is *Melipotis cellaris*.

ABOVE | Male White Witches (*Thysania agrippina*) have wingspans that sometimes reach 12 in (30 cm). They are strong flyers that are often seen out of range, especially in the fall. Adults are attracted to lights and fermenting baits.

NOCTUOIDEA: EREBIDAE: CALPINAE

FRUIT-PIERCING AND VAMPIRE MOTHS

Estimates of the diversity of the subfamily vary between 100 and over 900 species; recent treatments have removed genera and even tribes to other erebid taxa.

Core Calpinae include three mostly tropical and subtropical tribes: Calpini, Ophiderini, and Phyllodini. Many use the haustellum to pierce soft-skinned fruits; a few can even pierce the rinds of oranges and grapefruit to feed. Both ripe and fermenting fruits on the ground are targeted. Many will feed at tree wounds and other sugary, non-nectar resources.

The adults can be significant pests in orchards, as their feeding wounds can lead to fungal and bacterial infections that render the fruits unsellable. Vampire moths (*Calyptra*) employ their haustellum to pierce the skins of mammals to feed on blood!

Calpine caterpillars can be fantastic; a large number are rendered in bright yellows and oranges and black; others are snake mimics when threatened.

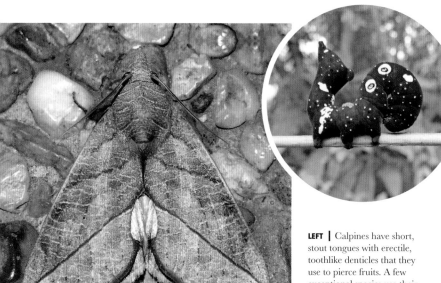

LEFT | Calpines have short, stout tongues with erectile, toothlike denticles that they use to pierce fruits. A few exceptional species use their tongues to take blood meals. Shown here is *Calyptra minuticornis*.

INSET | The function of the red "button" on A8 of *Eudocima homaena* has not been elucidated. It may draw attention away from the head—a bite targeting the button would allow the caterpillar to strike back.

DISTRIBUTION
Pantropical, with only modest representation beyond the subtropics

IMPORTANT GENERA
Calyptra, Eudocima, Gonodonta, and *Oraesia*

HABITAT
Tropical and temperate forests and woods

HOST ASSOCIATIONS
Specialists on woody plants; a few on herbaceous perennials

CHARACTERISTICS
• Medium-sized to large moths, many with yellow or orange HW; dorsal margin of FW often sinuate with projecting tooth before the tornus (visible in profile of resting adult)

• Stout haustellum with sharp denticles; upturned labial palpi

• Caterpillars varied; prolegs on A3 strongly reduced; those on A4 reduced in many genera; *Eudocima* serpent-like with prominent eyespots; *Gonodonta* aposematically colored

EPILOGUE: THE SIXTH EXTINCTION, NATURE, AND MOTHS

There could scarcely be a better or more urgent time to take up the study of moths and other invertebrates. Ours may be among the last generations to seek, discover, photograph, and study many of nature's creatures. We have already lost Sloane's Urania (*Urania sloanus*), and more than 300 species of Hawaiian moths have not been seen in the last 50 years—devastated by habitat loss, invasive species, introduced biological control agents, and now increasingly by climate change. Anthropogenic pressures are erasing the biota of the Atlantic palm forests of Brazil and Hispaniola, as well as the biota of the forests of Madagascar. Deforestation and desertification are devastating the biota of millions of acres of Africa each year. Globally climate change is overtaking habitat loss as the number one anthropogenic driver of biodiversity declines. We must do a better job of feeding the planet's 8 billion people in a more equitable manner than the generations before us. This will require the conversion of more, and still more, of Earth's forests, grasslands, and wildlands into croplands, exacerbating nature's challenges. Insects are the little things that run the world. There are many reasons why we should redouble efforts to appreciate, honor, and protect these creatures—for their ecosystem services and for the fates of our children, their children, and humanity—and actively embrace the stewardship of the planet's great bounty. The 10 million other species that share the planet with *Homo sapiens* have no vote, no seat at the table; they need our help so that they can continue their own struggles for survival and maintain their functions in the great web of life.

For those who study moths, the more they learn about them—the plants upon which they feed, their natural enemies, and other dimensions of biodiversity—the more they develop a greater appreciation, wonder, and desire to look after them. Moths are accessible animals in this regard and offer much in terms of what remains to be discovered: about their taxonomy, phylogeny, fossil history, life histories, population trends, behavior, biomechanics, interactions with other species, and conservation needs.

Beyond the pragmatic arguments for collecting, photographing, studying, and conserving moths, there is their ineffable beauty. They enrich our journeys. Hundreds are drop-dead gorgeous, as moths and as caterpillars, with some ranking among nature's most wondrous works. Although small in stature, with magnification, microlepidopterans reveal themselves to be worthy of much more attention. Heliodinids, heliozelids, choreutids, and so many others have resplendent forewing colors and patterns. Even unheralded moths, scaled in browns, mouse grays, and other earthen tones, can be fetching when examined closely. Fractals of moth wing patterns could surely be rendered into artisanal designs and stunning textiles.

It comes down to this: we live on a truly beautiful planet that is infinitely worthy of more study and stewardship. Moths make a wonderful gateway taxon to bring many people to this realization. As we learn about them, a new view of the world takes hold, along with a greater appreciation for the ecological functions they buttress, and the necessity of our collective efforts to protect our little-known planet.

GLOSSARY

Visit www.mothsoftheworld.com for an expanded glossary

Aedeagus: the male intromittent organ; phallus.

Aestivate: to pass the summer months in a state of inactivity; reproductive functions and behaviors are typically in stasis.

Androconia: male scent scales.

Angiosperm: a plant that produces flowers and bears its seeds in fruits.

Antenna: the anterior-most sensory organ for the detection of pheromones and other scents; also used for taste and touch, as well as for flight orientation.

Aposematic: boldly or warningly colored; also applicable to warning sounds.

Bet-hedging: adopting more than one strategy for survival or reproduction.

Biomass: the quantity (~weight) of organisms, living and dead, in a given volume.

Brachypterous: having shortened or rudimentary wings.

Bursa copulatrix: the pouch in the female abdomen that receives the spermatophore.

Chaetotaxy: the number, development, and position of larval setae; important in identification.

Chorion: the outer shell of an egg or ovum.

Chrysalis: an exposed pupa, free of a cocoon, as seen in Pterophoridae and sterrhines (Geometridae).

Clade: a lineage that includes all descendants of a shared common ancestor.

Congener: a member of a shared genus.

Costal: along the leading edge of the wing.

Coxa: the basal segment of an insect leg.

Crochets: the distal hooks on abdominal prolegs.

Crypsis: the stratagem of blending into the background to avoid detection.

Diapause: a period of hormonally controlled developmental and/or reproductive stasis.

Dicotyledonous: belonging to a dicot plant lineage; i.e., those with seeds that have two embryonic leaves.

Ditrysian: belonging to a lineage of moths with two reproductive openings, one for copulation and one for egg-laying.

Diurnal: day-active.

Dorsum: the top side of an insect.

Eclosion: the point of hatching from an egg or the emergence of an adult from a pupa.

Edaphic: having to do with soil.

Epiphysis: the flap of cuticle on the proleg used to clean the antenna.

Excrescences: minute integumental outgrowths (best viewed with land lens or other means of magnification).

Fauna: a set of animals that occupy a region.

Feculae: excreta, or caterpillar droppings; sometimes referred to as frass.

Femur: the third leg segment above the "knee."

Frenulum: the spine(s) from the base of the hindwing that is engaged by a finger-like lobe or set of curved scales on the underside of the forewing to couple the wings.

Frontal triangle: the frons, or triangular area, between the eyes on the head of a caterpillar.

Ganglion: a semi-autonomous neural mass that controls a segment or adjacent segments, independent of the brain.

Gymnosperm: a cone-bearing plant with seeds unprotected by ovary or fruit; compare with angiosperm.

Haustellum: the proboscis, or tongue.

Hemolymph: insect body fluid that, unlike blood, does not transport oxygen.

In copula: a coupled male and female in the process of mating, which is oftentimes extended in moths to allow for the transfer of the spermatophore.

Inquiline: species that lives in the nest, gall, or shelter of another.

Instar: the larval stage between molts; most moths have 4–7 instars, with 5 being the median; instars can differ significantly in form, duration, and habit.

Integument: the outer chitinous shell of an insect, shed at each molt.

Labium: the third and posterior-most set of mouthparts, bearing sensory palpi and a spinneret; labial palpi may extend forward, cover the face, or extend back over the thorax.

Larva: the second life stage of a moth; the principal feeding stage.

Macroheterocera: monophyletic group (clade) of five superfamilies: Drepanoidea, Lasiocampoidea, Bombycoidea, Geometroidea, and Noctuoidea.

Macrolepidoptera: historical, unnatural (polyphyletic) assemblage of Lepidoptera that included macroheterocerans, American moth-butterflies (Hedylidae), butterfly moths (Callidulidae), and butterflies; used in this guide in a much-restricted sense to refer to just the macroheterocerans moths, such that macrolepidopterans, macro moths, and macroleps can be used interchangeably.

Maxilla: the second and central set of insect mouthparts; always bearing sensory setae in larva and adult; portions of the maxilla form the haustellum.

Monophagous: feeding on plants of just one or two closely related families.

Monophyletic: an evolutionary group, derived from a common ancestor and that includes all its descendant lineages.

Monotrysia: a non-monophyletic grouping of pre-ditrysian moths that have a single reproductive opening for copulation and oviposition; five superfamilies are included: Adeloidea, Andesianoidea, Nepticuloidea, Palaephatoidea, and Tischerioidea.

Neonate: recently hatched larva

Nominate (taxon): the subgroup (taxon) upon which a higher category is based. For example, *Nola* is the nominate taxon of Nolinae as well as Nolidae.

Nocturnal: night-active.

Ocellus: a small, non-image-forming eye on the top of the head that plays roles in flight orientation and setting "biological clocks" of insects; there are two ocelli in moths that have them.

Pectinate: comblike; where branches (or rami) of antennae form a comb; moth antennae can be uni-, bi-, tri-, or quadripectinate.

Pheromone: a typically airborne liquid or solid chemical messenger used in communication between individuals of the same species.

Pinaculum (pl. pinacula): the cuticular base of a seta; often minute to small and round, but sometimes developed into a spine or horn.

Poikilotherm: a cold-blooded animal that does not maintain a constant body temperature.

Polyphagous: feeding on plants belonging to more than three families.

Proboscis: the haustellum, or tongue.

Pupa: the third stage of a moth, usually in a cocoon, soil, or wood; compare chrysalis.

Quadrifine venation: when the lower right corner of the discal cell in the hindwing appears to give rise to four veins (because the M2 vein is shifted downward and originates near that of M3).

Rami: extensions; the short branches from individual antennal segments.

Relict: a taxon or lineage that originated and diverged millions of years ago.

Retinaculum: the finger-like lobe or set of curved scales on the underside of the forewing that holds the frenulum, keeping wings joined in flight.

Scolus (pl. scoli): a stout, cuticular outgrowth usually bearing one to many setae.

Secondary seta: extra setae beyond the 11 pairs of setae common to most caterpillars; often added in second instar.

Sensillum: seta with sensory function in touch, smell, taste, hearing, carbon dioxide detection, gravity, moisture, and other environmental stimuli.

Seta: a hairlike outgrowth from the integument, ranging from fine hairs to bristles.

Spermatheca: an organ off of the common oviduct that stores and maintains sperm.

Spinneret: a spigot at the tip of the labium that secretes silk.

Spiracle: an opening on the side of the body for air exchange.

Stemma (pl. stemmata): a lateral, image-forming eye of caterpillars; most caterpillars have six stemmata on each side of the head.

Sympatric: co-occurring in the same geographic area.

Tarsus (pl. tarsi): the lower part of the leg that contacts substrate; composed of five tarsomeres; moths have tasting sensilla on their tarsi.

Taxon (pl. taxa): a rankless group or lineage that can apply to any grouping of organisms.

Tentiform: used to refer to the wing position in a perched moth, for taxa in which the wings descend steeply on each side of the body.

Tibia (pl. tibiae): the fourth segment of an insect leg, below the "knee," usually bearing one or two sets of spurs in moths.

Trochanter: the second, very small (scarcely visible), leg segment fused to the base of the femur.

Univoltine: having a life cycle with one generation per year.

Urticating: irritating to the skin; can induce a rash, with reaction sometimes severe.

Verruca (pl. verrucae): a fleshy wart bearing numerous (secondary) setae.

Vestigial: used to describe a structure of reduced complexity that has lost its ancestral function over evolutionary time.

ADDITIONAL READING AND IMPORTANT RESOURCES

WEBSITES

Addendum and Corrigendum for this Work: www.mothsoftheworld.com

AfroMoths: https://www.afromoths.net/

Australian Moths Online: https://www.csiro.au/en/research/animals/insects/id-resources/australian-moths-online

Barcode of Life Data System: https://boldsystems.org/

Biodiversity Heritage Library: https://www.biodiversitylibrary.org/ (resource for uncopyrighted taxonomic literature on all life forms)

HOSTS: Database of the World's Lepidopteran Hostplants: https://data.nhm.ac.uk/dataset/hosts

iNaturalist: https://www.inaturalist.org/observations

Leafminers of North America: https://charleyeiseman.com/leafminers/

Lepidoptera and Some Other Life Forms: https://www.funet.fi/pub/sci/bio/life/intro.html

Lepiforum: https://lepiforum.org/

Moth Photographers Group: https://mothphotographersgroup.msstate.edu/

Moths of Borneo: https://www.mothsofborneo.com/part-1/

Moths of India: https://www.mothsofindia.org/node/14

Moths of Japan: https://www.inaturalist.org/projects/moths-of-japan

Statement on Collecting of The Lepidopterists' Society: https://www.lepsoc.org/content/statement-collecting

UK Moths: https://ukmoths.org.uk/

BOOKS

Bernal, R., & B. Martinez (2023). *Polillas de Colombia: Guía de Campo* [Moths of Colombia: Field Guide]. Wildlife Conservation Society, Colombia Program.

Chacón, I., & J. Montero (2009). *Mariposas de Costa Rica* [Butterflies and Moths of Costa Rica]. Instituto Nacional de Biodiversidad, San Jose, Costa Rica.

Common, I. F. B. (1990). *Moths of Australia*. Melbourne University Publishing, Carlton, Victoria.

Gandy, M. (2016). *Moth*. Reaktion Books, London.

Inoue, H., S. Sugi, H. Kuroko, S. Moriuti, A. Kawabe, & M. Owada (1982). *Moths of Japan*. Vols. 1 & 2. Kodansha, Tokyo. (In Japanese.)

Kristensen, N. P. (ed.) (1998). Lepidoptera: Moths and Butterflies, Vol. 1: Evolution, Systematics and Biogeography. *Handbook of Zoology* IV/35: 51–63. Walter de Gruyter, Berlin & New York.

Lees, D., & A. Zilli (2021). *Moths: Their Biology, Diversity and Evolution*. Natural History Museum, London.

Manley, C. (2021). *British and Irish Moths: A Photographic Guide*, 3rd edition. Bloomsbury Naturalist. Bloomsbury Publishing, London.

Moths of Borneo (18-volume series led by Jeremy Holloway). Southdene, Iver, Buckinghamshire.

Moths of Thailand (7-volume series by different author sets). Brothers of St Gabriel in Thailand.

Powell, J. A., & P. A. Opler (2009). *Moths of Western North America*. Berkeley University Press, Berkeley, California.

Sourakov, A., & R. W. Chadd (2022). *The Lives of Moths: A Natural History of Our Planet's Moth Life*. Princeton University Press, Princeton, New Jersey.

Staude, H., M. Picker, & C. Griffiths (2023). *Southern African Moths and Their Caterpillars*. Pelagic Publishing, London.

Wagner, D. L. (2005). *Caterpillars of Eastern North America*. Princeton University Press, Princeton, New Jersey.

JOURNAL ARTICLES AND BOOK CHAPTERS

Kawahara, A. Y., *et al.* (2019). Phylogenomics reveals the evolutionary timing and pattern of butterflies and moths. *Proceedings of the National Academy of Sciences* 116(45): 22657–22663.

Mitter, C., D. R. Davis, & M. P. Cummings (2017). Phylogeny and evolution of Lepidoptera. *Annual Review of Entomology* 62: 265–283.

Rota, J., V. Twort, A. Chiocchio, C. Peña, C. W. Wheat, L. Kaila, & N. Wahlberg (2022). The unresolved phylogenomic tree of butterflies and moths (Lepidoptera): Assessing the potential causes and consequences. *Systematic Entomology* 47(4): 531–550.

van Nieukerken, E. J., *et al.* (2011). Order Lepidoptera Linnaeus, 1758. In Z.-Q. Zhang (ed.), Animal biodiversity: An outline of higher-level classification and survey of taxonomic richness. *Zootaxa* 3148(1): 212–221.

Wagner, D. L., E. Grames, M. L. Forister, M. R. Berenbaum, & D. Stopak (2021). Insect decline in the Anthropocene: Death by a thousand cuts. *Proceedings of the National Academy of Sciences* 118(2): e2023989118.

Wagner, D. L., & A. Hoyt (2022). On being a caterpillar: Structure, function, ecology, and behavior. In R. Marquis & S. Koptur (eds), *Caterpillars in the Middle: Tritrophic Interactions in a Changing World*. Springer, New York, pp. 11–62.

INDEX

Note: Superfamilies, families, and subfamilies are in Roman type; genera and species are in *italics*.

A

Abantiades hydrographus 91
Abantiades latipennis 90–91
Abbott's Sphinx 33
abdomen 24, 25, 30, 51
Acanthopteroctetes bimaculata 88
Acanthopteroctetidae 88
Acanthopteroctetoidea 88
Acentropinae 58, 160
Acharia stimulea 145
Acherontia atropos 33, 190
Acherontia lachesis 190
Acleris, A. variana 133
Acrolophus popeanella 106
Acronicta funeralis 28
Actias luna 6, 7
Actias selene 29
Adela 94
Adela reaumurella 95
Adelidae 43, 46, 94–95
Adeloidea 94–99
Aenetus virescens 91
Aetole bella 112
African Death's-head Sphinx 33, 190
Agathiphaga vitiensis 85
Agathiphagidae 18, 43, 46, 85
Agathiphagoidea 85
Aglossata 12
agricultural intensification, impact 72, 74, 231
agrochemicals 39, 72, 74, 75
Agrotis infusa 42, 54, 55, 213
Alcathoe caudata 137
Alcides 204
Alucita 122
Alucitidae 122
Alucitoidea 122
Amata compta 227
Ambulycini 191
American Swallowtail Moth 193
American Tent Caterpillar 171
Americerura scitiscripta 207
Amorpha 33
Amphipoea oculea 211
anatomy
  external 21–25
  internal 26
Andraca theae 177
androconia (scent scales) 16–17, 24, 36, 52, 53, 100, 154

angiosperms 9, 10, 15, 43
Anisota 184
Anomoeotidae 148
Antaeotricha schlaegeri 116
antennae 21–22, 23, 35, 36, 57
  larval 24, 25, 35
Anthela denticulata 176
anthelid moths 176
Anthelidae 176
Anticla antica 179
Antispila 96
ants 59, 71, 73
apatelodid moths 172–173
Apatelodidae 172–173
apophyses 35
Apoplania 89
aposematism 16, 56, 57, 62–64, 199, 222
aquatic lineages 58
aquatic snout moths 160
archaic bell moths 89
archaic moths 10, 12, 43, 56, 79, 85, 87, 102
  families/lineages 10, 12, 18, 43, 84–101
  mandibulate 84
archaic sun moths 88
Arctia caja 224
Arctic Woolly Bear Moth 30, 219
Arctiinae 217, 222–227
Arctiini 224–225
Arcyophora 209
Argema mittrei 65
Argent and Sable 201
Argiope 59
Army Cutworm 54, 213
armyworms 212–213
Arsenura 40, 185
arsenurid moths 185
Arsenurinae 185
Artichoke Plume Moth 123
artificial lights, threat to moths 72, 74–75
Athis inca 140
Atlas Moth 182, 183
Attacus atlas 182, 183
Australian Guava Moth 124
Autographa precationis 214
Automeris 57, 187

B

Bacillus thuringiensis 62
bagworm moths 29, 104–105
bait traps 70
baiting 68–70, 181
"balloon," by early instars 20, 104

ballooning 220
Barro Colorado Island, Panama Canal 73
Batesian mimicry 64, 137, 150
bats 13–14, 17, 19, 23, 34, 164
Beauvaria 61
behavior 49–64, 169–170
Bella Moth 45, 225
biological control 61–62, 73, 189
birds, caterpillars in diets for 15, 58, 59
Black-olive Caterpillar 209
Blackberry Skeletonizer 126
Blinded Sphinx 191
Bogong Moth 42, 54, 55, 213
Bombycidae 178–180
Bombycoidea 24, 172–192
bombykol 179
Bombyx mandarina 179, 180
Bombyx mori 37, 38, 172, 178, 180
Box Tree Moth 158
Brahmaeidae 172, 175
Brahmin moths 175
brain 26
Brenthia 130, 131
bristle-legged moths 126
bristles/bristlelike spines 23, 24, 25, 88, 107, 126, 127
broods 30–31
Brown-tail Moth 219
Bryolymnia viridata 19
Bucculatricidae 46, 110
Bucculatrix coronatella 110
buck moths 50, 186-187
Buff-tip Moth 59
Bumelia Webworm Moth 128
burnet moths 16, 149–150, 151
bursa copulatrix 49, 50, 51, 52, 102
butterfly moths 140

C

Cactoblastis cactorum 157
Calliduliidae 154
Calliduloidea 154
calling females 36, 50, 51, 52, 169
Caloptilia 108
Calpinae 230
Calyptra minuticornis 230
camouflage *see* mimicry
Campaea perlata 199
Canephora hirsuta 105

Carmenta 137
Carpenterworm Moth 138
carpet moths 200–201
Carposina sasakii 124
Carposina scirrhosella 124
Carposinidae 124
Carposinoidea 124
Carthaeidae 177
Case-bearing Clothes Moth 40, 107
case-bearing moths 117
Castniidae 56, 140
Catapterix 88
caterpillars 6, 15, 28–29
  anatomy 16, 24–25, 26
  crypsis *see* mimicry
  diets, plant specialization 35, 43, 44–45, 218
  form/color changes 28, 29
  hairy 25, 172, 174, 219, 221
  as human food 40–42, 59, 182, 183
  hunting/collecting 65–66, 70, 78
  instar number 28
  internal feeders 25, 46, 88
  overwintering 31
  parasitized 60, 61
  as pests *see* pests, moths as
  predation of 15, 58, 59–60
  silk use by 20, 25, 105, 168–171, 180
  sound production 33, 190, 192
  toxic compound sequestration 44, 58, 62, 225
  unpalatability 44, 58, 63, 142, 172, 174, 204
Catocala 68, 69, 228, 229
Catocala fraxini 228, 229
catocalines 228
Cauchas fibulella 95
Cecropia Moth 22
Cephonodes hylas 192
Ceratocampinae 184
Cerodirphia candida 187
Ceromitia 95
chaetotaxy 25
Chalcosiinae, and chalcosiines 149
Chelepteryx collesi 176
chemoreceptors 34, 35
Chiasmia clathrata 199
Chihuahuan Desert, moths 31
Choreutidae 130–131
Choreutoidea 130–131

235

chorion 27
Choristoneura fumiferana 134
chrysalises 30
Chrysiridia 204
Chrysiridia rhipheus 6, 204
Chrysopsyche lutulenta 168
Cicinnus melsheimeri 163
Cimeliidae 166
circulatory system 26
Citheronia regalis 181
clades 13, 14, 23, 172
classification of moths 8, 12, 14
clearwing moths 23, 136–137
Clifden Nonpareil 228, 229
climate change 72, 74, 75, 231
clothes moths 39–40, 107
cocoons 20, 30, 37–38, 110, 180
Cocytius antaeus 18
Codling Moth 135
Coleophora albicosta 117
Coleophora alnifoliae 117
Coleophoridae 46, 117
collecting moths 68, 77–78, 79, 231
collections, insect/moth 77–78
color 8, 16, 56, 231
  aposematism 16, 56, 57, 62–64, 199, 222
  caterpillars 28, 29, 44, 58, 63
  diurnal *vs* nocturnal moths 56, 57, 94
  eggs and pupae 27, 30
color vision 32, 188
Coloradia 186
Coloradia pandora 186
Comadia redtenbacheri 40, 41, 139
Comet Orchid 48
Composia credula 64
Composia fidelissima 222
Compsilura cocinnata 61
conservation 28, 71–78, 79, 231
Copiopteryx 185
Coptodisca 96
Cordyceps 61, 62
corkscrew moth 193
Corn Earworm 38
coronet moths 48
Coscinoptycha improbana 124
cosmet moths 119
Cosmopterigidae 119
Cosmopterix 119
Cosmopterix zieglerella 119
Cosmosoma 58
Cossidae 138–139
Cossoidea 136–140
Cotton Bollworm 39, 215
courtship 23, 33, 49–50, 51
courtship brushes 24, 53
crambid snout moths 158

Crambidae 13, 46, 56, 158–161, 159, 160, 161
Crambinae 158, 159
Crambus praefectellus 159
Creatonotos gangis 223
crochets 25, 164
crypsis, larval *see* mimicry
cryptic moth species 24, 33, 44–45, 62, 90
Cryptophasa rubescens 113
Cryptothelea gloverii 29
cutworm moths 210–211, 212–213
Cyana 226
Cyana malayensis 226
cyanide 16, 62, 150
Cyclophora 202
Cyclosia papilionaris 150
Cydalima perspectalis 158
Cydia pomonella 135
Cydia saltitans 135

D

Dacnonypha (infraorder) 12
Dalceridae 143
dart moths 212–213
Dasychira 220
Deborrea malgassa 105
defensive chemicals 16, 17, 57, 62–63, 224, 225
  pyrrolizidine alkaloids 44, 45, 58, 62, 225
deforestation 72, 74, 231
Depressariidae 115, 116
Depressariinae 115
deserts/drylands 31
Diamondback Moth 39
diapause 31, 49, 66
dichomeridines 121
Didugua argentilinea 63
distribution 57–58, 73
Ditrysia (ditrysian moths) 11, 13, 49, 102–103
diurnal moths 17, 18, 56–57, 83, 94
  finding/photographing 69–70
diversity of moths 8–9, 13, 45, 57–58, 65
Doidae 166
Domesticated Silk Moth 37, 38, 172, 178, 180
Doratifera quadriguttata 145
Douglas moths 125
Douglasiidae 125
Douglasioidea 125
Drepana 167
Drepanidae 164, 166–167
Drepanoidea 164, 166–167
droughts 74, 75
ductus seminalis 102
dusk, finding moths 66
Dysodia 152

E

Ear Moth 211
Earias 209

ears 13, 18–19, 23, 34, 69, 203
echolocation, bats 34
ecology 57–58
economic importance of moths 37–42
  see also food (for humans); pests, moths as; silk
ecosystems, insect decline 71, 73
Ectoedemia atricollis 93
education, about moths 75, 76, 231
eggs 27–28, 31, 51, 59, 60, 102
Eichlinia cucurbitae 139
Elachistidae 67
emerald moths 196–197
emperor silk moths 181–182, 183
Enaemia pupula 142
endangered moth species 42, 55, 71, 78
Endotricha ignealis 156
Endoxyla 139
Endoxyla leucomochla 40
Endromidae 177
Endromis versicolora 177
Ennominae 198–199
Entomophaga maimaiga 62
Eois 200, 201
Epermenia chaerophyllella 127
Epermeniidae 127
Epermenioidea 127
Ephestia kuehniella 39
Epia muscosa 179
Epicampoptera 167
Epipleminae 205
Epipyropidae 141
Epitymbiini 134
erebid moths 216–217
Erebidae 13, 56, 64, 69, 216–230
Erebinae 217
Eriocraniidae 87
Eriocranoidea 87
Euclea obliqua 145
Eudesmia arida 46
Eudocima homaena 230
Eudryas grata 210
Eulepidotis 217
Eupackardia calleta 58
Eupithecia 200, 201
Eupithecia absinthiata 200
Euproctis 220
Euproctis chrysorrhoea 219
Eupterote mollifera 174
Eupterotidae 172, 174
European Corn Borer 161
Euryglottis aper 57
Euxoa auxiliaris 54, 213
evolutionary origin, moths 10–14, 15, 16, 18–19, 81, 82, 85

plant coevolution 43, 44, 45
Exoporia (infraorder) 12
extinct species 71, 231
eyed sphinx moths 191
eyes, compound 21, 22, 32, 56, 57
eyespots 57, 160, 183, 187, 191, 195, 203, 205

F

Fabiola 114
facultative broods 31
fairy moths 43, 56, 94–95
Faithful Beauty 222
Fall Armyworm 39
false burnet moths 128
families of moths 13
Fangalabola moth 105
females
  antennae 21–22
  control of courtship 51
  pheromones 35, 36, 49, 50, 52, 59
  reproductive anatomy 51
Fiery Clearwing 36
finding moths 65–67
flannel moths 146–147
flat-bodied moths 115
flies, parasitoid 60, 61
flower(s), for moths 47, 48, 70, 75
flower moths (Noctuidae) 29, 215
food (for humans) 40–42
  caterpillars as 40–42, 59, 182, 183
  moths as 42, 186
  pupae as 41, 105
food webs 15, 37, 164
footman moths 46, 226
forelegs 23
Forest Tent Caterpillar 169, 170
foretibia 23
forewing 21, 23–24, 79
fossils (moth) 10, 81
frenulum 23, 24, 88
fringe-tufted moths 127
fruit-piercing moths 230
fruitworm moths 124
Funerary Dagger 28
fungi infecting moths 41, 42, 61–62
fungus moths 106–107

G

galacticid moths 132
Galacticidae 132
Galacticoidea 132
gall forming 46, 108–109
ganglia 26
Garden Lance-wing 127
Garden Tiger 224
Garella nilotica 209
Gelechiidae 120–121

236 Index

Gelechioidea 113–121
genitalia 24, 49, 50, 51, 52
geographic distribution 73
geological history 10–14
geometrid moths 194–195
Geometridae/geometrids 13, 20, 24, 27, 34, 56, 194–202
Geometrinae 196–197
Geometroidea 13, 164, 193–205
ghost moths (Hepialidae) 12, 18, 24, 27, 41, 42, 46, 66, 90–91
Giant Peacock 183
giant silk moths 36, 40, 67, 181–182
Giant Sphinx 18
global threats to insects 72, 73, 231
glory moths 177
Glossata 12
Gloveria 169, 170
glycols 31
goat moths 138–139
Gonimbrasia belina 40, 42, 183
Gracillariidae 46, 108–109
Gracillariinae 109
Gracillarioidea 108–110
Grapholita molesta 135
grass-miner moths 67
grass snout moths 159
Greater Death's-head Sphinx 190
Gum Leaf Skeletonizer 209
Gynaephora groenlandica 30, 219

H

habitat 57–58, 74, 76
Hadena 48
Hadenini 212
Hag Moth 145
hairy body (moth) 168, 169
hairy caterpillars 145, 146, 176
handmaiden moths 227
haustellum (proboscis) 12, 17–18, 21, 22, 34, 57
Hawaii 71, 126, 231
hawk moths 32, 57, 188–189
head, anatomy 21–22, 24, 25
Helicoverpa 215
Helicoverpa armigera 39, 215
Helicoverpa zea 38
Heliocheilus 215
Heliodinidae 112
Heliolonche 215
Heliothinae 29, 215
Heliothis 215
Heliozelidae 96
Hemaris 48
Hemaris aethra 189
Hemerophila 131
Hemerophila diva 131
Hemileuca 181, 186

Hemileuca grotei 187
Hemileuca hera 187
Hemileuca maia 50
Hemileucinae 182, 186–187
hemolymph 26, 63, 199, 225
Hepialidae (ghost moths) 12, 18, 24, 27, 41, 42, 46, 66, 90–91
Hepialoidea 90–91
Hepialus humuli 90
Herald Moth 49
herbivory, toxic compounds to reduce 35, 43–44, 150
Herminiinae 218
hermit moths 113
Heterobathmia pseuderiocrania 86
Heterobathmiidae 43, 86
Heterobathmiina 12
Heterobathmioidea 86
Heterocera 82, 164
Heteroneura (infraorder) 12, 13, 14
Hilarographa 133
Himantopteridae 148
hindlegs 23
hindwing 21, 22, 23–24, 79
histamines 221
holometabolous development 16–17
Homadaula anisocentra 132
hooktip moths 166–167
hornworms 189
Hyalophora cecropia 22
Hyblaea puera 153
Hyblaeidae 153
Hyblaeoidea 153
Hyles euphorbiae 189
Hyles gallii 188
Hyles lineata 65
Hylesia 186
Hylesia metabus 63, 64
hypermetamorphic development 28

I

Idaea 27, 202
Idaea degeneraria 202
identification of species 75
Imma 129
immid moths 129
Immidae 129
Immoidea 129
inchworms 195
Incurvaria masculella 97
Incurvariidae 97
Indian Moon Moth 29
Indianmeal Moth 39
insect decline 72–73
insect walks 76
intoxication of moths 69
invertebrate predators of moths 59
io moths 186–187
Iridopsis 195
IUCN, imperiled species 71, 73

J

jawed moths 10
jewel caterpillar moths 143
Jordanita chloros 150
Jumping Bean Moth 135
jumping-spider mimic 10, 84, 112, 114, 118, 130, 131, 160

K

Kauri moths 85
Kentish Glory 177
keratin, digestion 40, 45, 106–107
kill traps 68, 78
knot-horned moths 157

L

labial palpi 21, 22, 25, 34, 35, 189, 192
labium 21
Lacosoma 162
Lacosoma arizonicum 163
Lacosoma chiridota 163
Lacturidae 142
Lactura sapotearum 142
Lampronia 97
Lampronia corticella 99
Lanassa lignicolor 13
land-use change 72, 74
Langiinae 188
lappet moths 168–171
Larentiinae 200–201
larva/larval stage 16, 28–29, 30
  see also caterpillars
Lasiocampidae 164, 168–171
Lasiocampoidea 10, 168–171
leaf blotch miner moths 108–109
leaf-mining 46, 47, 86, 87, 88, 93, 95, 101, 109
leafcutter moths 97
legs 22, 23
Lepidoptera 8, 12, 15
  evolutionary origin 10, 81, 82, 102
lichen-feeding caterpillars 19
lichen moths 223, 226
life cycle 27–31
light detectors, ocelli as 21, 32, 33
light pollution 72, 74–75
light sources attracting moths 9, 32, 67–68, 79
light traps 67–68, 224
Limacodidae 144–145
Lithocolletinae 109
Lithophane leautieri 212
Lithosiini 46, 226
litter moths 218
live-trapping methods 67–68, 78
Lobster Caterpillar 207
Lomoymia 187

long-tailed burnet moths 148
Lonomia 186, 187
looper moths 194–195
Luna Moth 6, 7
Lymantria dispar 38, 61, 62, 221 220
Lymantriinae 219
Lyssa 203

M

Macalla thyrsisalis 155
Macaria carbonaria 17
Macaria liturata 198
macro moths 164–230
Macroglossinae 188, 192
macroglossine sphinx moths 192
Macroheterocera 164–230
macrolepidoptera 11, 13, 34, 164–230
Madagascan Moon Moth 65
Madagascan Sunset Moth 204
Maguey Worm 40, 41, 139
Malacosoma americanum 171
Malacosoma disstria 169, 170
  males 102–103
  antennae 21
  pheromones 36, 52
  sexual conflict 51–52
mandibles 21, 25
mandibulate archaic moths 84
Manduca 190
Manduca blackburni 78
Manduca sexta 189
many-plumed moths 122
Maroga melanostigma 113
mass migration 53, 54–55, 204
mating 49–50, 50–51, 51–52, 53, 102, 178
  Utetheisa 225
mating swarm 59
maxilla 21, 25
maxillary palp 21, 22, 25, 35
Mediterranean Flour Moth 39
Megalopyge lanata 146, 147
Megalopygidae 146–147
Meganola 208
Melipotis 228, 229
Melipotis cellaris 229
mesothorax 23
Metalectra diabolica 46, 216
metalmark moths 130–131
metamorphosis 15–17, 27–31
metathorax 23
microlepidoptera 10, 27, 30, 54, 81–100
Micronia 203
Micropterigidae 10, 43, 83
Micropterigoidea 83
Micropterix 83

migration 53–56, 65, 204
Milionia zonea 194
Mimallonidae 10, 162–163
Mimallonoidea 162–163, 164
mimicry 10, 57, 62–64, 84, 150, 226
  by caterpillars 19–20, 44, 172, 192, 193, 198, 199, 201, 230
Mimosa Webworm 132
Mnesarchaea acuta 92
Mnesarchaeid moths 92
Mnesarchaeidae 92
Mnesarchaeoidea 92
Mnesarchella 92
molting 26, 33
Mompha propinquella 118
Mompha raschkiella 118
momphid moths 118
Momphidae 46, 118
monkey moths 174
Monoleica semifascia 145
Monopis dorsistrigella 107
Monotrysia 49
Mopane Worm 40, 42, 183
moth gardens 70, 75
mouthparts 16, 17
Müllerian mimicry 58, 64, 150, 226
mummies (of Hepialidae) 41–42

N

natural enemies of moths 58–62
nectar/nectaring, coilable tongue for 17–18
Nemophora 94
Nemophora staudingerella 95
Nemoria 197
Nemoria arizonaria 197
Neopseustidae 89
Neopseustis meyricki 89
Neopseustoidea 89
Nepticulidae 18, 46, 93
Nepticuloidea 93
nervous system 26
nests, lasiocampid (tent caterpillars) 20, 169–170, 171
Nisiga simplex 174
Noctuidae 13, 19, 29, 38–39, 46, 48, 56, 164, 165, 210–215
Noctuinae 46
Noctuoidea 13, 23, 36, 164, 206–230
nocturnal moths/nocturnality 18–19, 56, 67, 68
Nola 208
Nola cucullatella 209
Nolidae 208–209
Notodontidae 206–207
nursery pollination systems 48

O

oak trees 38, 58, 110, 221
observation, of moths 65–70
ocelli 21, 22, 32
Oecophora bractella 114
Oecophorid moths 114
Oecophoridae 114
Oiketicus 105
Olceclostera seraphica 172
Old World butterfly-moths 154
Olethreutinae 135
olethreutine leafroller moths 135
olfaction 21, 36
oligophagous species 44
ommatidia 32
Operophtera 201
Opisthoxia 195
Oreta rosea 166
Orgyia 220
Orgyia antiqua 28
Oriental Fruit Moth 135
Oriental Silk Moth 37, 38, 172, 178, 180
Orthosia 213
ostium bursae 51
Ostrinia nubilalis 161
overwintering 31, 122, 213
ovipositor 35, 102
owl moths 175
owlet moths 34, 46, 210–211

P

Palaephatid moths 100
Palaephatidae 100
Palaephatoidea 100
Panacelinae 174
Pandora moths 186–187
Paonias excaecata 191
Paracrama dulcissima 208
Parasa indeterminea 7, 145
parasites 60
parasitoid wasps 60–61
Parategeticula 98, 99
Parectopa lespedezaefoliella 109
pathogens 61–62
Peach Fruit Moth 124
Pectinophora gossypiella 121
Pennisetia marginata 136
Perola clara 145
pesticides/agrochemicals 39, 72, 74, 75
pests, moths as 38–40, 132, 201, 211, 213, 214, 221
  of clothes/textiles 39–40, 107
  of crops 38–39, 53, 121, 122, 123, 124, 127, 135, 137, 159, 189
  of forests/trees 134, 139, 153, 182
  of stored-products 39, 156, 157
petite leafmining moths 93
Phalera 207

Phalera bucephala 59
Pharmacis 41–42
Phazaca 205
phenotypic plasticity 197
pheromones 24, 35, 36, 49, 50, 52–53, 59, 154, 178
Phiditiidae 177
Philobota 114
Phobetron pithecium 145
photography 69, 70, 75, 79, 231
Phragmataecia castaneae 139
Phycitinae 157
Phyllocnistinae 109
Phyllocnistis populiella 47
Phyllonorycter leucographella 109
phylogeny of moths 10–14, 56, 81–82, 103, 165
Phymatopus hectoides 52
Pindi Moth 90–91
Pink Bollworm 121
Plagodis alcoolaria 19
Plagodis dolabraria 198
plant-feeding guilds 45–46
planthopper parasite moths 141
plants, and moths 9, 43–48, 70
  coevolution 43, 44, 45
  diversity, moth diversity 9, 45, 58
  larval diets, host-plant specialists 35, 43, 44–45, 218
  pollination 18, 32, 43, 46–49, 70
  secondary plant compounds 35, 43–44, 45, 171
Platyptilia carduidactyla 123
Plodia interpunctella 39
plume moths 30, 66, 123
Plusiinae 214
Plutella xylostella 39
Polka-dot Wasp Moth 227
pollinators, moths as 18, 32, 43, 46–49, 70
polyphagous species 44, 223
pre-pupa 29, 31
predation, of caterpillars/moths 15, 58, 59–60
Primrose Moth 215
Prionoxystus robiniae 138
proboscis (tongue) 21
  see also haustellum
processionary moths 206–207
Procridinae 149
Prodoxidae/prodoxids 29, 43, 98
Prodoxus 31, 99
Prodoxus decipiens 99
Prodoxus y-inversus 99
prolegs 24, 25, 173
prominent moths 206–207
prothorax 22
Prothysana 172
Psilopygida 184

Psychidae 104–105
Pterodecta felderi 154
Pterolocera 176
Pterophoridae 30, 66, 123
Pterophoroidea 123
Pterophorus pentadactyla 123
"puddling" 18, 53, 54
pupae/pupal stage 29–30, 31, 36, 59, 61, 105
pyralid snout moths 155, 156
Pyralidae 155–157
Pyralinae 156
Pyralis 156
Pyraloidea 13, 24, 34, 155–161
Pyraustinae 158, 161
Pyromorpha dimidiata 150
Pyropteron chrysidiformis 36
pyrrolizidine alkaloid (PA) 44, 45, 58, 62, 225

Q

quakers 213

R

Raspberry Moth 99
Rattlebox moths 225
Reed Leopard 139
reproductive openings, two 13, 49, 51, 102
respiration 26
retinaculum 23, 24
retronecine 44
rewilding 76
Rheumaptera hastata 201
Richia albicosta 47
ribbed cocoon-maker moths 110
root feeding 46, 47
royal moths 184

S

Sabatinca kristenseni 10
sack-bearer moths 162–163
salivary-silk gland 25
satin moths 219
Saturnia pyri 183
Saturniidae 40, 41, 49, 67, 181–187
Saturniinae 183
scales 8, 16, 17, 22
Schinia 215
Schinia florida 215
Schreckensteinia festaliella 126
Schreckensteiniidae 126
Schreckensteinioidea 126
scientific collections 77
scoli (horns) 58, 175, 182, 184
Scoliopteryx libatrix 49
scoopwing moths 203, 205
Scopula 202
Scorched Wing 198
secondary plant compounds 35, 43–44, 171
seeds, moths feeding on 19, 29, 48

segments 21, 22, 24, 25, 30
Sematuridae 193
semi-looper moths 214
senses 32–36
sensilla 21, 25, 34, 35
Sesiidae 36, 46, 46, 56, 136–137
setae 23, 25, 34, 35, 87, 168, 170, 174
  primary, secondary 25, 146, 169, 175
  stinging/toxin-laden 144, 146, 219, 220, 221
sexual conflict 50–52
shapeshifters 28
shield-bearer moths 96
Shorea 148
silk 20, 25, 104, 111, 180
  economic importance 37–38, 180
  production/use by caterpillar 20, 25, 105, 168–171, 180
silk worms (silk moths) 20, 25, 41, 178–180
Sinna calospila 209
Sloane's Urania 71, 204, 231
slug caterpillar moths 144–145
smell and olfaction 21, 36, 47, 179
Smerinthinae 188, 191
Smerinthini 191
smoky moths 62, 149–150
sound production 33–34, 190, 192, 208
sound reception 13, 19, 34, 164, 168, 192
South America, moth species number 8–9, 58
spanworms 195
Sparganothini 134
species of moths, number 8, 9, 13, 79
sperm 49, 51, 102
spermatophores 49–50, 51–52, 102, 225
Sphecodina abbottii 33
Sphingidae/sphingids 32, 48, 164, 188–192
Sphinginae 188, 190
Sphingognatha asclepiades 174
sphinx moths 33, 34, 48, 188–189, 190
spiders 59, 60, 225
Spilomelinae 161
spinneret 25
Spodoptera frugiperda 39
Spodoptera litura 39
Spongy Moth 38, 61, 62, 220, 221
Spruce Budworm 134
Squash Vine Borer 135
Stauropus alternus 207
stemmata 32–33
Stenomatinae 116

stenomatine moths 116
Sterrhinae 30, 202
Stigmella castaneaefoliella 93
stinging setae/spines 144, 146, 219, 220, 221
Stiria intermixta 19
Stiria rugifrons 210
sugary baits 68–70
sun moths 112, 140
sunset moths 18, 54, 203, 204
superfamilies of moths 13, 79, 82, 102
Synemon jcaria 140
Syntomeida epilais 227
Syntomini 227
Syssphinx blanchardi 184
Syssphinx raspa 43

T
tails 66, 148, 185
Taro Caterpillar 39
tarsi 22, 34
taste/tasting 34–35
Tawny-barred Angle 198
taxonomic names 14
Teak Defoliator 153
teak moths 153
Tegeticula 98, 99
Tegeticula yuccasella 98
tent caterpillars 20, 168–171
terminalia 50–52
Thitarodes 41–42
thorax 21, 22–23
  caterpillars 24, 25
Thyrididae 152
Thyridoidea 152
Thysania agrippina 229
Tiger Moth 61, 63
tiger moths 33–34, 44, 217, 222–223, 224–225
timber moths 113
Timocratica 116
Tinagma 125
Tinagma ocnerostomella 125
Tinea 107
Tinea pellionella 40
Tineidae 39–40, 45, 106–107
Tineoidea 104–107
Tischeriidae 46, 101
Tischerioidea 101
tongue, coilable/coiled 12, 17–18, 21, 22, 34, 57
Tortricidae 13, 46, 56, 133, 134, 135
Tortricinae 134
tortricine leafroller moths 134
Tortricoidea 133–135
Tortyra slossonia 131
toxic alkaloids 34
toxic chemicals *see* defensive chemicals
trees 9, 38, 58, 68
Trisyntopa 114
tropical burnet moths 142
tropical regions 53, 54, 55, 57–58

Trosia 146, 147
trumpet leafminer moths 101
tufted moths 208–209
tussock moths 219
twirler moths 120–121
tymbal 208

U
ultrasonic sounds 33–34
  from bats, detection 13, 19, 34, 164, 168, 192
ultraviolet light, moths attracted by 9, 32, 67
underwing moths 69, 216–217, 228–229
unpalatability of caterpillars 44, 58, 63, 142, 172, 174, 204
unpalatability of moths 57, 62, 63, 178
  color and 16, 44, 56, 57, 58, 62–64, 222
  toxins/chemicals *see* defensive chemicals
Uraba lugens 209
Urania fulgens 53
Urania leilus 204
Urania sloanus 71, 204, 231
Urania Swallowtail Moth 53
Uraniidae 54, 203–205
Uraniinae 204
Urodidae 29, 128
Urodoidea 128
Urodus parvula 128
Utetheisa 225
Utetheisa ornatrix 45, 225

V
vampire moths 230
Vamuna remelana 223
Vapourer Moth 28
viruses, entomopathogenic 62
vision 32–33, 57, 188
voucher specimens 77

W
Wallace's Sphinx Moth 18, 48, 188
wasp moths 58, 227
wasps, predation by 59, 60–61
wave moths 202
webworm caterpillars 20
webworm snout moths 161
Western Bean Cutworm 47
White-lined Sphinx 65
White Plume Moth 123
White Witches 229
Wild Silk Moth 179, 180
wild silk moths 183
window-winged moths 152
wings 16–17, 22–23, 85
  coupling mechanism 23–24
winter moths 201, 212, 213
witch moths 228

Witchetty Grub 40, 139
Wood Leopard Moth 139

X
Xanthopan praedicta 18, 48, 188
Xyloryctidae 113

Y
Yellowstone National Park 54–55
Yponomeuta 111
Yponomeutidae 111
Yponomeutoidea 111–112
yucca moths 31, 43, 48, 98–99

Z
Zeugloptera 12, 84
Zeuzera pyrina 139
zombie caterpillar, protecting wasps 62
Zygaena 16, 149, 151
Zygaena trifolii 151
Zygaenidae 56, 62, 149–150
Zygaeninae 149
Zygaenoidea 141–151

ACKNOWLEDGMENTS

Niels Peder Kristensen's (1998) unrivaled volume on Lepidoptera for the *Handbook of Zoology* series anchors this effort, as do those that authored the family treatments in Kristsenen's volume. Markku Savela's website on Lepidoptera proved especially valuable, as was Gaden Robinson *et al.*'s HOST database. I owe much gratitude to Ken-ichi Ueda and other programmers, patrons, and the legion of contributors who have collectively built iNaturalist—it is an invaluable resource. The people that contributed images greatly elevated the aesthetics and other aspects of this work and are hereby thanked; the many images shared by Piotr Naskrecki and Mike Thomas warrant special recognition. David Price-Goodfellow played integral roles in image selection, editing, and layout. Richard Brown, Sue Carnahan, David Dussourd, David Lees, Ryan St Laurent, Tanner Matson, and Alberto Zilli made invaluable contributions to the text. Alexandra Thornton and Madeline Shaw drafted the maps; Alexandra also assisted with the phylogenies. Many colleagues read and provided feedback on the taxon accounts: John Brown (Tortricoidea), Richard Brown (Tortricidae), Nick Dowdy (Introduction, Arctiinae), Marc Epstein (Zygaenoidea), Axel Hausmann (Geometridae), Kevin Keegan (Noctuoidea), Bernard Landry (Pyraloidea), David Lees (archaic families), Tanner Matson (Geometroidea), Richard Peigler (Bombycoidea), Jadranka Rota (Introduction, Choreutidae), Brian Scholtens (Pyraloidea), Ryan St Laurent (Mimallonidae, Notodontidae, and Bombycoidea), Jim Tuttle (Sphingidae), and Alberto Zilli (Noctuoidea). Leslie Goethals assisted with acquisition of many images from iNaturalist. Alexandra Thornton, Oliver Malatich, and Owen Brown checked taxonomic spellings. Financial support came from the Richard P. Garmany Fund, US Forest Service (Department of Agriculture), Earthwatch, National Science Foundation (DEB 1557086, 2225092), and Connecticut Department of Environmental Protection.

PICTURE CREDITS

The publisher would like to thank the following for permission to reproduce copyright material and for use of reference material for illustrations. All reasonable efforts have been made to contact copyright holders and to obtain their permission for the use of copyright material. The publisher apologizes for any errors or omissions and will gratefully incorporate any corrections in future reprints if notified.

@ferox1003 212r. **Alamy Stock Photo**: Al Argueta 74; Auscape International Pty Ltd 55; Zoltan Bagosi 199t; Judith Bicking 20; BIOSPHOTO 36; Biosphoto/Frank Deschandol & Philippe Sabine 175c, 225l; Biosphoto/Jean-Yves Grospas 204r; blickwinkel/H. Bellmann/F. Hecker 105r; blickwinkel/Lenke 115c; blickwinkel/R. Sturm 155t; Joerg Boethling 41t; Ron S. Buskirk 69; Nigel Cattlin 39, 107tr; Phil Degginger 185; Larry Doherty 94l, 198l, 212l; F1online digitale Bildagentur GmbH/Matthias Lenke 32l; fishHook Photography 8b; Galaxiid 30; Jeffrey Isaac Greenberg 18+ 76; David Havel 57l, 57r; imageBROKER.com GmbH & Co. KG/Hans Lang 33l; INTERFOTO 37c; Jeff Lepore 197c, 199l; Rainer Lesniewski 42t; mauritius images GmbH/Solvin Zankl 77; Charles Melton 56r; Minden Pictures/Michael & Patricia Fogden 98; Minden Pictures/Steve Gettle 147c; Mint Images Limited 73; Nature Picture Library/John Abbott 210l; Nature Picture Library/Paul Williams 134r; James Peake 176l; Ian Redding 211; Bryan Reynolds 213c; Dinal Samarasinghe 60t; Malcolm Schuyl 204l; Sunshower Shots 105l; Tom Tookey 201; James Weber 18t; Wirestock, Inc. 43t. **Giff Beaton** 45r, 145ucl, 181l, 210r. **Michał Brzeziński** 177t. **Lyle J. Buss**, University of Florida 128r. **Chris Conlan** 187tcl. **Geoffrey Cox**, South Australia, April 2022 100. **Rob Curtis** 137t. **Nicolas J. Dowdy** 23. **Charley Eiseman** 93r. **Glenn Fine** 186. **Flickr**: Mike Budd/USFWS 66t; Christina Butler 159; Lior Carlson 68; Cataloging Nature 179t, 200l, 203l; Patrick Clement 12b, 28tl, 84l, 87t, 87b, 93l, 94r, 97l, 99b, 101l, 101r, 108, 109l, 114, 117l, 117r, 118t, 118b, 125t, 127l, 135r, 167b, 206, 209l; Bernard Dupont 64t, 155l, 173t, 208r, 209br, 217, 229b; Antonio Giudici 148; Nick Goodrum 224l; Janet Graham 96r, 161l; Donald Hobern 113r, 144, 156, 160, 176r; Jason Hollinger 47b; Jean and Fred Hort 91t, 140t, 145lcl; Tanaka Juuyoh 95t; Katya 139; Pavel Kirillov 179b, 227; Dean Morley 29tl, 224r; Andy Reago & Chrissy McClarren 166l, 216r; Line Sabroe 200r; Katja Schulz 116, 136, 137b, 45tl, 150l, 216l; Adam Searcy 53; LiCheng Shih 89r, 129t, 141l, 194r, 207t, 207uc; Bernard Spragg 182r; Forest & Kim Starr 78; Michael Taylor 17; Ilia Ustyantsev 115tr, 120, 121c, 134l, 138, 150c, 188; Dinesh Valke 174l; Nicolas Venner 20; Wayne National Forest 35c; Len Worthington 190, 230l; Alexey Yakovlev 63b, 192t; Sarah Zukoff 47t. **Getty Images/thietkebep02 /500px** 56l. **George Gibbs** 10, 84l. **Leslie Goethals** 67r, 70. **iNaturalist**: Criss Acuña 172b; Amanda Stasse Barrantes 143r; Brandon Boyd 222; Logan J. Bradley 169; Robert Briggs 209tr; Jacy Chen 177b; Christine 170b; Floro Ortiz Contreras 29cl; Juan Cruzado Cortés 229l; Vinicius S. Domingues 184l; douglaseustonbrown 104; Lin Sun Fong 153l, 205; CheongWeei Gan 174r; Laura Gaudette 162; Ellyne Geurts 187tl; Jacqui Geux 91c; Ferhat Gundogdu 152t; Patrick Hanly 170tl; Takaaki Hattori 179c; Nick Helme 95b; Jody 132r; Ben Keen 213t; Michael Knapp 125b; Michel Langeveld 119l; lisemari 111l; Diogo Luiz 141r; Dan MacNeal 133cr; Chrissy McClarren and Andy Reago 31, 106, 132l; Jake McCumber 215; Michelle 187tr; Jana Miller 29ctr; Sascha Nun 111r; Grete Pasch 18b; Cathy Powers 113l; Liliana Ramirez-Freire 140b; Katja Schulz 110r, 119r; Sterling Sheehy 129c; SK53 127r; Erin Springinotic 170tr; Kai Squires 147t; Christopher Stephens 124l; Richard Stovall 112; Suncana 122b; sunnyjosef 154l; tjeales 122t, 142t; Tobyyy 213b; Simon Tonge 59l; Julien Touroult 187cr; Ken-ichi Ueda 143l; Barry Walter 198r; Robert Webster 110l; Eridan Xharahi 131b; Mathew Zappa 218; K. Zyskowski & Y. Bereshpolova 193. **Samuel Jaffe**, The Caterpillar Lab 214r. **R. Stephen Krotzer** 221. **Suipoon Kwan** 153r. **Alan Chin Lee** 80. **Claudio Maureira** 86. **Tom Murray** 107b, 133t. **Piotr Naskrecki** 2, 5, 168. **Nature Picture Library/Robert Thompson** 181r. **M.W. Nelson/MassWildlife** 50. **John D. Palting** 9. **Jim Petranka** 99t. **David Price-Goodfellow** 16, 158, 183, 207b. **Buck Richardson** 223r. **Chris Rorabaugh** 131t. **Jadranka Rota**, Lund University Biological Museum, Sweden 25. **Gilles San Martin**, Walloon Agricultural Research Centre (CRA-W) 97r. **Dave Sargeant** 150tr. **Science Photo Library**: Dennis Kunkel Microscopy 12t; Eye of Science 22bc. Steve Gschmeissner 34; Susumu Nishinaga 35r; Martin Oeggerli 8cr; David Scharf 32c. **James Schroeder** 54. **Thomas Shahan** 130. **Shutterstock**: Noel V. Baebler 171; Karel Bock 65; Davide Bonora 192c; chaypunn 180; Stefan Dinse 59r; Catherine Eckert 157r; Fawwaz Media 60c; J.J. Gouin 189cl; Habitante 41b; Elliotte Rusty Harold 195b; David Havel 66b; Mark Heighes 228; JorgeOrtiz_1976 40, 107t; Tomasz Klejdysz 161r; Henri Koskinen 90; Nikolay Kurzenko 7; Hugh Lansdown 71; meechai39 38; Young Swee Ming 61b; Sandra Moraes 146; sarin nana 178; Matee Nuserm 149r; IanRedding 126t; Simon Shim 223t, 226; Milkov Vladislav 189cr; W. de Vries 15l; HWall 151l, 167t, 220; winaryanti 133cl; weerapong worranam 37b; YoONSpY 22bl. **Leroy Simon** 214l. **Christopher Stephens** 92. **Andrei Sourakov** 223l, 225r. **Mark S. Szantyr** 33r. **Ryan S Terrill** 64b. **Michael Thomas** 45l, 48, 58c, 58b, 61c, 147b, 166r, 187tcr, 187br, 189t, 191, 195t, 196, 207bc. **Catalina Tong** 149l. **David L. Wagner** 6, 11,13, 19t, 19c, 19b, 28cl, 28cr, 43b, 46t, 46c, 52, 62l, 62r, 63t, 72, 88t, 88c, 91b, 109r, 126, 128l, 131c, 135l, 138t, 142c, 145r, 145ucr, 145lcr, 145br, 151r, 152c, 155r, 163t, 163c, 163b, 172r, 173b, 182c, 184r, 197t, 219c. **Len Willan** 85. **Christine Young** 96l. **Wiki Commons**: Adrian Tync Gower 49; Jeevan Jose, Kerala, India 203r; Muséum de Toulouse MHNT/Didier Descouens 219r; Plate from William Swainson *Zoological illustrations*, Volume 3, 2nd series. *Leilus Occidentalis* = *Papilio Sloaneus* Cr. Accepted as *Urania sloanus* (extinct), 1829 71; Luekhope Tsuchoi 2000 42c; Vinayaraj 230r. **Wikidata**: USDA Agricultural Research Service/Peggy Greb 121t, 157l. **Wikipedia**: Gyorgy Csoka, Hungary Forest Research Institute, Bugwood.org 124r; Charles Lam 154r; Ivar Leidus 123.